P9-CQV-002

INDIAN RIVER CO. MAIN LIBRARY

Encyclopedia of
Sustainability

Encyclopedia of Sustainability

Business and Economics
VOLUME II

Robin Morris Collin

Robert William Collin

Greenwood Press
An Imprint of ABC-CLIO, LLC

Santa Barbara, California • Denver, Colorado • Oxford, England

Library of Congress Cataloging-in-Publication Data

Encyclopedia of sustainability / Robin Morris Collin, Robert William Collin.
3 v. cm.
Includes bibliographical references and index.
Contents: vol. 1. Environment and ecology —
ISBN 978-0-313-35263-8 (vol. 1 print : alk. paper) — ISBN 978-0-313-35264-5 (vol. 1 e-book) — ISBN 978-0-313-35265-2 (vol. 2 print : alk. paper) — ISBN 978-0-313-35266-9 (vol. 2 e-book) — ISBN 978-0-313-35267-6 (vol. 3 print : alk. paper) — ISBN 978-0-313-35268-3 (vol. 3 e-book) — ISBN 978-0-313-35261-4 (set - print : alk. paper) — ISBN 978-0-313-35262-1 (set - e-book)
1. Environmental sciences—Encyclopedias. 2. Sustainability—Encyclopedias. 3. Sustainable development—Encyclopedias. I. Collin, Robin Morris. II. Collin, Robert W., 1957–
GE10.E528 2010
333.7203—dc22 2009037029

14 13 12 11 10 1 2 3 4 5

This book is also available on the World Wide Web as an eBook.
Visit www.abc-clio.com for details.

ABC-CLIO, LLC
130 Cremona Drive, P.O. Box 1911
Santa Barbara, California 93116–1911

This book is printed on acid-free paper ∞
Manufactured in the United States of America

Copyright Acknowledgments

The authors and publisher gratefully acknowledge permission for use of the following material: Appendix B: Courtesy Equator Principles Financial Institutions, Work Ethics Ltd, www.equator-priniciples.com; Appendix C: The Ceres Principles. Source: Ceres & Investor Network on Climate Risk. Reprinted with permission.

Contents

VOLUME TWO: BUSINESS AND ECONOMICS

Guide to Related Topics

Preface

References to sustainability are everywhere, from advertising to space travel. Words associated with sustainability are fast becoming ubiquitous. This reference text is designed to help understand the many meanings of sustainability.

Concepts of sustainability have been developed in multiple disciplines including the sciences, international agreements, development law and policy, and humanities. The essential concepts about sustainability can be described in terms of three broad domains:

- Environment and ecology
- Business and economics
- Equity and fairness

The relationship between these domains is described in somewhat different ways. Some describe their relationship as a three-legged stool or three intersecting circles. Each circle or leg of the stool represents one domain, environment, economics, and equity. Each circle is of equal size; each leg bears equal weight.

Others describe the fundamental relationships as three nested baskets. The environment is the largest, most comprehensive basket. Within it, all human activity is located including human economic enterprises and human communities. Economic enterprise is represented by a basket nested within our environment and its webs of life. Human individuals

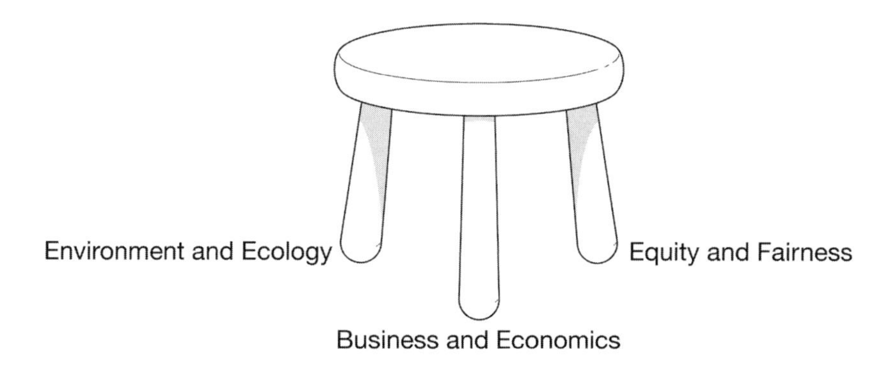

Figure 2.1 • Sustainability as a three-legged stool: Environment, economics, and equity. Illustrator: Jeff Dixon.

and their communities rest within both these two baskets relying on each for their livelihood and support. This encyclopedia set will provide the reader with the necessary infrastructure to navigate the complex roads and byways of the contemporary discourse on sustainability. That infrastructure is based on the axes of environment and ecology, economics and business, and equity or fairness. Sustainability weights these three areas equally and joins them in every discourse. This triangulation of the so-called three "E"s—environment, economics, and equity—distinguishes sustainability as a philosophy different from that of conservationism or environmentalism. This encyclopedia devotes one volume to each of the three "E"s.

In each volume, there is the same basic organization of chapters: a comprehensive introduction, definitions in contexts that are pertinent to that particular volume, the contemporary public policy contexts arranged from the global to national to local levels, current controversies, and future trends.

Each volume begins with a comprehensive overview of what the term *sustainability* means in each domain. This comprehensive overview introduces the major concepts of sustainability as used in each unique

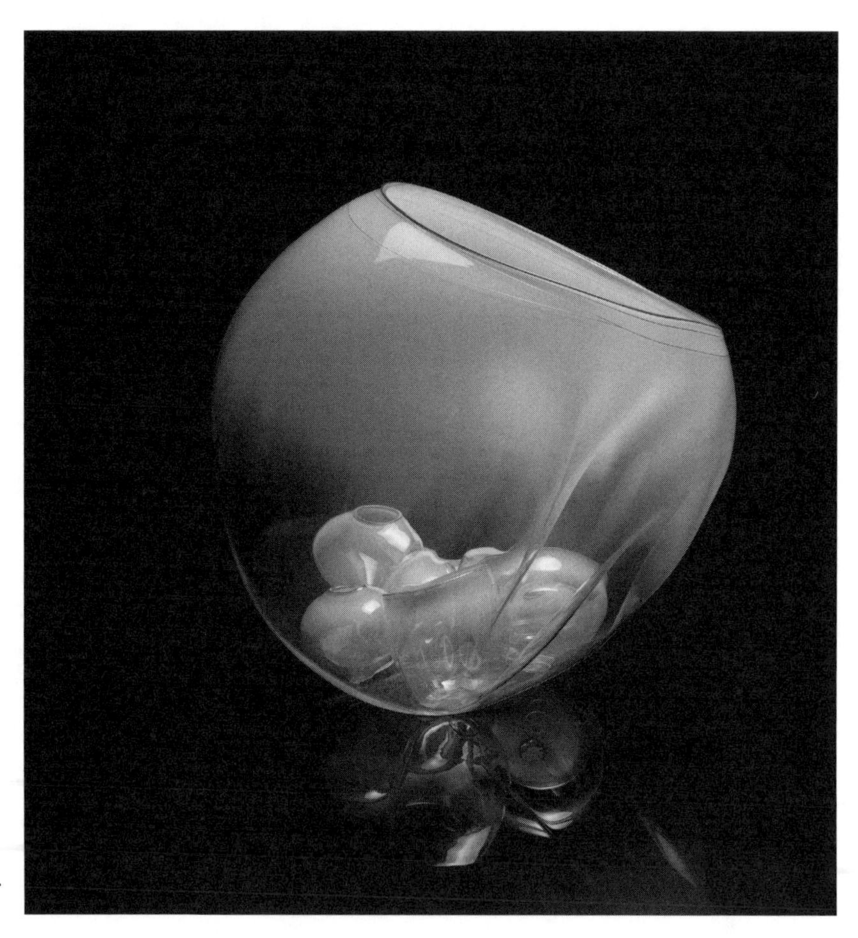

FIGURE 2.2 • Sustainability as three nested baskets: Environment, economics, and equity. Copyright © Dale Chihuly, 1992. Sky Blue Basket Set with Cobalt Lip Wraps, 1992 17″ × 15″ × 16″. Photo by Terry Rishel.

volume: environment and ecology, economics and business, and equity or fairness. This introductory chapter provides a concentrated account of the "big picture" in each unique arena of sustainability. The interconnectedness of basic ideas makes the study of sustainability challenging to the novice and complex to anyone. The overview section of each volume presents basic concepts and relationships in the primary context of the volume. The overview provides concise insight into terms of art, the dynamics that have shaped the concept within that context, and the changes and challenges that sustainability presents in that particular context. These are the elements of complex interactions and the background on which human choices and policies act and interact with natural systems.

Following the overview, each volume contains a chapter on definitions and contexts that give in-depth descriptions of key terms set into contexts relevant to that particular area. This chapter is a primer on the basic terms and definitions that are foundational to each volume. Without such a primer, terms of art can become a private language of expertise, making knowledge and information inaccessible to the general public reader, even one with considerable education. This section identifies key terms of art and defines them in accessible language. Definitions are arranged contextually, and alphabetically. This organization is tailored to assisting the reader with quick and ready access to the language and contexts of the sustainability discourse. This chapter gives a reader ready access to the specialized terms of art and foundational definitions of the discourse in each area. The dynamic interaction of these terms is illustrated in images and examples throughout this volume.

Next in each volume Chapter 3 describes the role of government and the United Nations in achieving sustainability. We include a description of United Nations programs while recognizing that the UN is not a government but an association of sovereign governments. The UN has exercised global leadership in guiding world governments toward sustainability. We also describe the work of national, regional, and local governments. Local governments have a uniquely important role to play in implementing sustainability because they are most closely connected to place and community. Even in countries whose national governments have chosen to abdicate their role in achieving sustainability, local governments have acted independently as laboratories and as activist organizations to achieve important behavioral changes. Many United Nations programs and policies are aimed at this local and regional level of government as well as through nongovernmental organizations (NGOs).

In the government section of each book, public policy is arranged in a global to local progression and, within that structure, it is presented in its chronological order. We have organized each chapter on government involvement to trace the historical developments as they have occurred at different levels: global developments through the United Nations organization, U.S. national developments where they have contributed, and the state, local and regional efforts within the United States. Sustainability public policy is rapidly unfolding in these different venues at uneven rates of change. For example, U.S. municipalities are rapidly adopting the Precautionary Principle in land use ordinances designed around sustainability.

In Chapter 4 of each volume, controversies related to sustainability are explored in greater depth. Every environmental and developmental conflict reflects competing and sometimes conflicting interests of many interested parties. Too often, these are portrayed in the popular media as simple conflicts between two parties. The truth about controversies, however, requires an appreciation of the competing and conflicting interests of multiple stakeholders. In the end, all solutions will be local, as are all environmental and ecological controversies. Solutions are not the focus of this section of the volumes here. Instead, we aim to describe the nature of the interests involved in all aspects. Controversies are once again arranged alphabetically. Each volume lists and describes current controversies in each volume. These controversies were selected based upon their political salience, and likely impact upon future generations. Controversies around sustainability provide a rich area for classroom discourse.

Finally, each volume concludes with a section devoted to emerging trends arranged alphabetically. This chapter takes a considered look at trends and data to follow for future developments. These trends were selected partially based upon the availability of existing data collections, and the commitment of governmental organizations and nongovernmental organizations to collect and monitor data. New sources of data and better resources will continue to develop but these fundamental trends should lead the interested reader to the sources that we have now. In this section, we describe what the future of sustainability in each area may hold based on the information available now. At this moment, some changes seem inevitable, whereas others may be subject to human management. The future of our relationships built on a sustainable model of dynamic and radical inclusion is the subject of imagination and possibilities.

The concept of sustainability is divided into three co-equal components of environment, economy and equity. We did so to arrange the large amount of information in a manner most comprehensible to the reader. These components of sustainability are dynamic as well as interrelated. We developed our framework of overview, definitions, government involvement, controversies, and future trends per volume to help the reader understand sustainability in the context of each of the co-equal components. Within each framework, we have further introduced concepts of scale, such as going from global to local levels of government intervention. When appropriate, we have introduced chronologies that underscore the development of sustainability. Each volume is unique, and capable of standing alone, but with a similar framework. Each volume is cross-referenced, with portal Web sites. Our goal is to provide a comprehensive framework that the reader can easily navigate within and between volumes.

The ultimate challenge of human sustainability is how human enterprise and communities that can establish themselves without undermining the fundamental health of the natural systems on which all beings on Earth rely. The legacy of previous centuries, with their development of human enterprises and communities of great scale reliant on diminishing ecological resources, is the presence of wealth amidst poverty and environmental degradation that challenges our ability to survive. Sustainability as a doctrine challenges us to provide for our needs while allowing future generations the same opportunities for prosperity and a full experience of life. The idea that we can do that for future generations while ignoring the growing inequities of contemporary life is an equal challenge to sustainability. The central challenge of sustainability is how human enterprises and communities can function within ecosystems supporting our environment. The great human progress and development of the contemporary era have been accompanied by a growing gap between rich and poor people in the context of widespread environmental deterioration, and increasing poverty. Earth's ecosystems provide enormous benefits for humankind. Some of those benefits are from renewable resources, and some are not renewable. Humankind has exceeded the limits of some renewable resources and is approaching the limits of nonrenewable ones.

For those who are interested in the idea of sustainability and wish to explore it further, there is an overwhelming volume of material to read devoted to specific contexts and applications. Often, this material is difficult to penetrate for a novice because it is so heavily reliant on specialized language and specialized constructs unique to a particular discipline. Theses volumes provide a gateway that allows access to the full variety of the field and facilitates independent investigation in a multidisciplinary field. Sustainability will require new ways of thinking about the environment and a basic shift in public policy at all levels of government. This reference is dedicated to the task of facilitating human imagination and thought in that direction. Imagination is a uniquely human faculty praised in physics and metaphysics alike. Albert Einstein said that imagination was more important than information. Buddhism insists that thought and intention are as important as the acts to which they may give birth. Creativity may be inspired by the interplay between the major perspectives offered in each volume. Pragmatic implementation is illustrated in stories, biographies, and illustrations throughout these volumes. These are designed to help the reader understand key concepts, and thereby provide a springboard for the next generation of human imagination and sustainability.

References

Anderson, William. 2001. *Economics, Equity, Environment.* Washington, DC: Environmental Law Institute.

Capra, Fritjof. 1996. *The Web of Life: A New Scientific Understanding of Living Systems.* New York: Anchor Books.

Collin, Robert William. 2007. *Battleground: Environment.* Westport, CT: Greenwood Press.

Collin, Robin Morris, and Robert William. "Where Did All the Blue Skies Go? Sustainability and Equity: The New Paradigm." *Journal of Environmental Law and Litigation* 9 (1994):399–460.

Dubash, Novraz K., and Daniel Bouille. 2002. *Power Politics: Equity and Environment in Electricity Reform*. Washington, DC: World Resources Institute.

Johnson, Steven M. 2004. *Economics, Equity and the Environment*. Washington, DC: Environmental Law Institute.

Paehlke, Robert. 2008. *Democracy's Dilemma: Environment, Social Equity, and the Global Economy.* Cambridge, MA: MIT Press.

ACKNOWLEDGMENTS

We would like to express our gratitude for all those who helped us with this encyclopedia. Willamette University's President Lee Pelton and the Center for Sustainable Communities provided foundational support. We are also grateful to David Paige at ABC–CLIO Press, for his support, patience, and timely assistance. We are grateful to the students who worked as research assistants for us in this process: Sikina Hasham, Lacey Lucas, and Sarah Hunt Vasche. Candace Bolen provided invaluable office support for us for which we are grateful. We would also like to thank all our students over the years. More than a decade of teaching the first sustainability course in a U.S. law school, several sustainability courses in university environmental programs in the United States and abroad, and environmental justice issues to communities and government agencies has exposed us to a wonderful cohort of earnest, thoughtful, and hopeful students. These students are from Auckland University, New Zealand; Cambridge University, UK; Urban and Regional Planning, Jackson State University; Department of Urban and Environmental Planning, University of Virginia; Environmental Studies and Law schools at the University of Oregon; Willamette, Tulane, and Lewis and Clarke Law Schools, Hunter College, and Cleveland State University School of Social Work. Many communities and indigenous peoples have also shared their visions of sustainability with us. We are very grateful for their assistance. They include the Choctaw Band, the Spokane Tribe, and the Indigenous Environmental Network. Many community groups and their leaders have shared their hopes for justice and sustainability. They include the Environmental Justice Advisory group, Center for Community Environmental Justice, and many urban community leaders. The federal and state environmental agencies also have their share of students we are grateful for. The U.S. Environmental Protection Agency, the Oregon Department of Environmental Quality, the Washington Department of Ecology, and the Missouri Department of Environmental Quality all contributed students of sustainability. All of our students provide us with a window to the future, a future they want to be sustainable.

CHAPTER 1

Introduction and Overview

This overview presents a quick explanation of sustainability dynamics as they relate to the business and economic basis of sustainability. This explanation is necessary to understand the next section on business and economic definitions and contexts for sustainability. These concepts are part of the emerging language of business and economics sustainability. We begin with the ecological systems on which all human enterprises are founded.

WEBS OF LIFE: THE INTERRELATEDNESS OF NATURAL SYSTEMS

All life on Earth is supported by its biological, chemical, and physical processes. Sustainability in the context of economics and business recognizes that all livelihoods have their foundations in the systems that support life on Earth. Sustainable economics and business recognize that our economies have compromised the ability of these systems to provide for future generations when they are undervalued. The challenge to sustainability in this context is to find new ways to make livelihoods, to eliminate waste and pollution in what we make and consume, and, perhaps most significantly, to change our thinking about our needs, wants, and desires—in short, what it is that makes for human happiness.

Many authors and teachers are urging us to think about our economies as "nested" within our ecosystems. These systems configure the elements of nature in ways that support all living things. Ecosystems refer to constellations of shared conditions that support life in a local area. A particular area on Earth may share water resources, biological species, climatic conditions, and cultural traditions that interact to support distinctively adapted local life. The elements of an ecosystem are dynamic processes themselves, and they interact with each other. For example, within a watershed, processes of evaporation, transpiration, and condensation link land, air, and water.

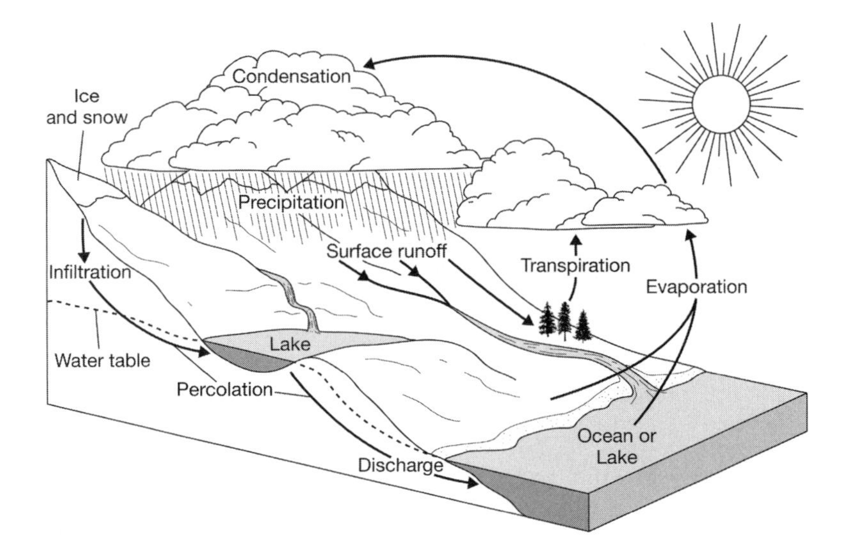

FIGURE 2.3 • Water cycle: The interaction of land, air, and water in a watershed. Illustrator: Jeff Dixon.

All human development comes from these basic ecosystems and the services they provide. All growth is supplied by them. Human development has successfully interacted with ecological functions to meet human needs. We have done this through agriculture by plowing and tilling soil and redistributing water for the growth of plants for food, fiber, and fodder. Hunter-gatherer societies take other species as prey to meet their need for food and fiber. Human uses of ecological functions have grown in scale in response to population growth and because of technology. Human beings have thrived and increased in numbers and complexity of social arrangement because of our successful use and exploitation of these natural systems. The ability of these systems to produce a flow of goods for human use, however, does not tell us about the health of these environmental systems.

Concerns about the health of ecosystems and the life processes that support them have led scientists to study the baseline conditions of Earth's life support systems. From 2001 to 2005, 1,360 scientists from around the world worked together to assess the world's ecosystems and the services they provide to human development. Their report, called the Millennium Ecosystem Assessment (MA), identified 10 categories: marine, coastal, inland water, forest, dry land, island, mountain, polar, cultivated, and urban. These are not ecosystems in and of themselves; rather, they are each composed of ecosystems. Each of these MA categories shares similar biological, climatic, and social factors that make them distinctive. The ecosystems within each category offer benefits to humans living there. These are referred to in the MA as ecosystem services. Ecosystem services are the benefits people obtain from ecosystems. These services are divided into four broad groups: provisioning resources, regulating services, cultural services, and supporting services.

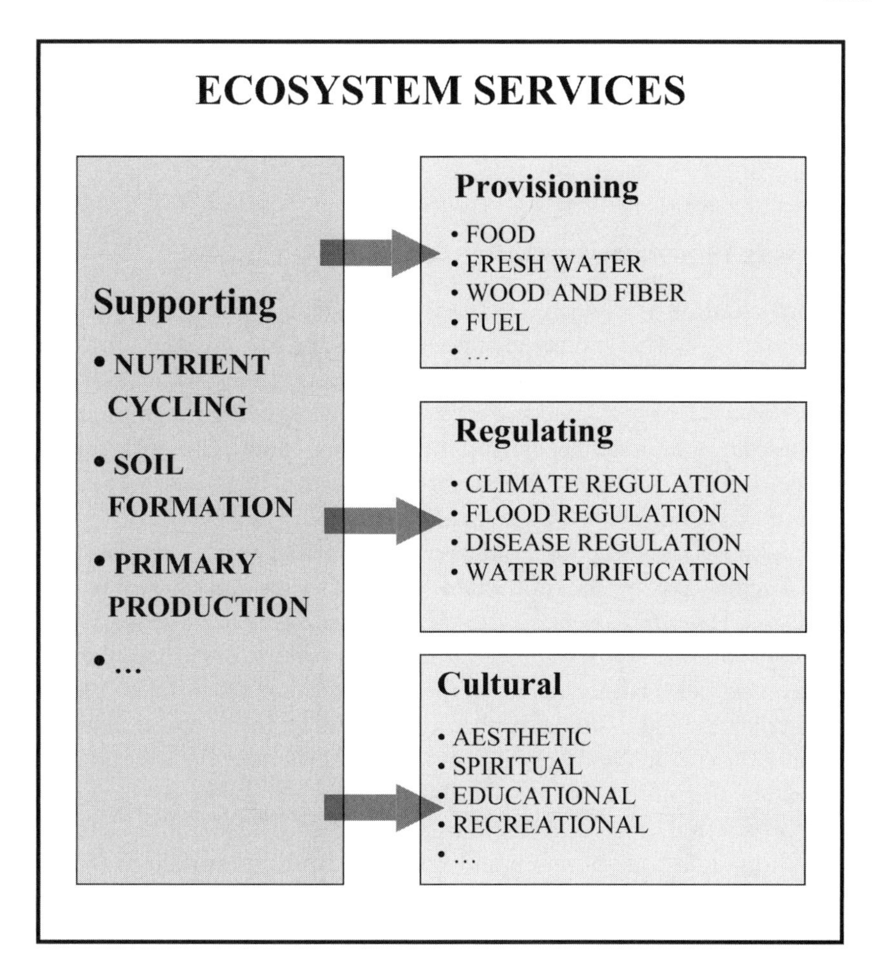

Figure 2.4 • Ecosystem services are the benefits that people obtain from ecosystems. Millennium Ecosystem Assessment • www.maweb.org.

HUMAN IMPACTS ON ECOSYSTEMS

When human activity engages an ecosystem, it changes that system, as does any life form. All economic activity engages our ecosystems to obtain material resources and energy. Business has a tremendous effect on the human communities in which they are present, as well as on ecosystems. These effects include employment rates, health conditions at workplaces, and health impacts of business operations outside the workplace. Profits alone do not reflect significant costs of production in these areas. Some companies have embraced responsibility for these ideas by accounting in ways that go further than financial profits. Corporations that use a "triple bottom line" approach account for profits, as well as environmental and social costs. The scale of the human activity, however, affects ecological impacts. ***See also* Volume 2, Chapter 2: Language of Sustainable Business.**

References

Egan, Michael. 2009. *Barry Commoner and the Science of Survival: The Remaking of American Environmentalism*. Cambridge, MA: MIT Press.

Jacobs, Jane. 1992. *Systems of Survival: A Dialogue on the Moral Foundations of Commerce and Politics.* New York: Vintage Books.

Savitz, Andrew W., and Karl Webber. 2006. *The Triple Bottom Line: How Today's Best Run Companies Are Achieving Economic, Social, and Environmental Success.* Hoboken, NJ: Jossey-Bass.

What Is Industrialization?

Industrialization is one type of social arrangement organizing the use of technology. The industrial model of production rapidly consumes resources and energy to produce consumer goods. In industrialized societies, this model has framed society's view of ecosystems as commodities and resources rather than life support systems. Industrialization frames human labor as a cost of business rather than part of human individual identity and dignity. Industrialization transforms production of goods from a household-by-household basis into a factory model of specialization. Factory-type production often separates the product from the web of life and ecosystems necessary to create it, again creating a view of consumer products as items of profit rather than products of natural systems. Farming is still a family-operated model of production in many places, with cultivation and animal husbandry embedded in the social arrangements of family life and community cultural structures. When the scale of cultivation and husbandry is embedded in a factory model of industrial production, it becomes agribusiness or a contained animal feeding operation. The scale of waste and pollution from industrialized modes of production is much greater. For example, the U.S. Greenhouse Gas inventory estimates that emissions from industry contributed 20 percent to overall U.S. greenhouse gas emissions in 2007, agriculture contributed another approximately 7 percent, and the commercial sector accounted for about 6 percent. The residential sector contributed about 5 percent.

Governments of the industrial age embraced the technologies and social arrangements necessary to create an industrialized economy. The policies of that age facilitated the construction of necessary infrastructure for transportation and labor to produce goods on a scale never before possible. Dramatic consequences came from these policies, especially in the generation of wealth for industrialized countries that often procured the resources necessary for advancement—natural resources and labor resources—by conquest and colonial practices. Other consequences include improvements in public health, increased population, widespread pollution, and poverty, especially in communities and among people victimized by conquest. Sustainability forces us to engage these unresolved controversies as they reemerge in the context of sustainable development discussed in Chapter 4.

Technology and Industrialization

Technological innovation has accelerated human development and growth. Technologically driven impacts on our planet's ecological systems are both intentional and unconscious. The internal combustion engines made it possible to farm vast tracts of land and connect communities at great distances. The result is more food for more communities. It also releases greenhouse gases that cause climate change and discharges materials that injure human health, and its basic production and transportation cause contamination of land and oceans. The greatest human impact on ecological functions in both scale and quality has come from industrialization.

GLOBALIZATION: DEVELOPMENT OR DISASTER?

Globalization refers to the ever-increasing network of connections between individuals and organizations made possible by technology and transportation. These networks include information technologies, as well as other communications that facilitate the transfer of capital and goods, and redistribute labor around the globe. Globalization can mean ever-widening markets, as governments agree to remove barriers to trade and immigration between them. Globalization can also mean the ubiquitous presence of markets through virtual technologies. In earlier centuries, markets were limited in time and locations, as well as by social class. Today, market transactions can occur virtually anytime and anywhere.

Markets have been good vehicles for increasing wealth, but growth in marketplaces has also accompanied alarming trends in the rate of climate change and other environmental deterioration, as well as growth in poverty. Businesses are not often required to do accurate environmental impact statements, or otherwise account completely for their actual or cumulative ecosystem impacts. Market models that do not incorporate actual or cumulative impacts are not compatible with most models of environment or ecosystem based sustainability. Black markets, such as illegal markets for endangered animals, are by definition unregulated by any government. The effect of black market transactions on ecosystems is unknown.

Communications have increased exponentially using satellite and microwave technologies together with digitization of data. Computers have made management of these technologies possible on a grand scale. These communications technologies by themselves take a toll on resource and energy use. The average laptop uses 45 tons of raw materials and is rarely recycled. Many of these technologies use energy even when they are not in use, prompting the name "energy vampires." These technologies have also been used for business and commercial purposes. Capital is readily transferable using these communications technologies. Producers now sell to a variety of buyers that were not reachable in the ordinary sales transaction without communications technology and computers. This has increased demand for products. In the production of new products, raw materials can be purchased and transported for assembly and resale globally. The availability of transportation has also made labor itself a global commodity, with businesses moving toward the least costs of production including cheap labor. Traditional governmental restrictions on the movement of goods between marketplaces have been removed through trade agreements. In these ways, the marketplace for goods and services has been greatly expanded. The resultant growth in markets and transnational corporations has dwarfed many national economies.

Transportation is also global and much quicker than ever before. The dependence of transportation companies on oil and gas, however, has made this network unsustainable. Assumptions around the cost of transportation and the ability to transport goods may affect the scale of globalization that we see in the future. Climate change may affect global transportation links as rising ocean levels force relocation of ports and reveal polar routes through previously non-navigable waters.

Globalization is a complex idea that encompasses the integration of economic, political, and cultural systems across the globe. There are conflicting ideas about whether globalization is a positive force that includes economic growth, prosperity, and ideas about democratic freedom or whether it includes change that results in environmental devastation, exploitation of the developing world, and suppression of human rights. That the global atmosphere has changed because of human activities is now an established fact.

There has been a recent increase in the pace and scope of contact and ties between human societies across the world. The globe and its inhabitants are becoming increasingly interdependent and connected. This has been spurred by technological advances in communication and transportation. National borders are increasingly porous, allowing a diverse group of people to interact. Because of the effects of globalization, ideas and cultures circulate more freely. This mixing of ideas can have a positive impact; however, these changes are also uprooting old ways of life and threatening cultures that have previously been insulated from outside impact.

Globalization creates new markets and wealth. These new markets are often very lucrative for some people in the world, but not for others. For example, many medicines come from the tropical rainforests. Indigenous peoples' knowledge is critical to understanding the uses of these plants. Pharmaceutical companies have patented indigenous knowledge of medicine, making billions of dollars in profits without including the indigenous people whose knowledge contributed to the understanding of these substances. Globalization is a powerful, ambivalent force. It can bring immense change, often benefiting those most in need, but sometimes devastating their lives and livelihoods. As the term implies, globalization touches the lives of people all over the globe.

Municipalities and local governments are taking a more active role in collaborating with local business to shape the footprint and trajectory of materials used in their communities. This is especially true in the United States, where the national level of government has withdrawn support for UN agreements on climate change and has not pursued the recommendations of the 1999 Presidential Council on Sustainable Development, Towards a Sustainable America: Advancing Prosperity, Opportunity, and a Healthy Environment for the 21st Century, May 1999, PCSD's Final Report to the President. ***See also* Volume 2, Chapter 2: Globalization; Volume 2, Chapter 3: United States.**

References

Binley, Dan, and Oleg Menyailo. 2004. *Tree Species Effects on Soils: Implications for Global Change.* New York: Springer.

Clapp, Jennifer, and Peter Dauvergne. 2005. *Paths to a Green World: The Political Economy of the Global Environment.* Cambridge, MA: MIT Press.

Hess, David J. 2007. *Alternative Pathways in Science and Industry: Activism, Innovation, and the Environment in an Era of Globalization.* Cambridge, MA: MIT Press.

Mitchell, Ronald B., William C. Clark, David W. Cash, and Nancy M. Dickson. 2006. *Global Environmental Assessments: Information and Influence.* Cambridge, MA: MIT Press.

World Bank. 2008. *The Global Monitoring Report.* Washington, DC: World Bank.

Values of Industry

By contrast, the values of industrialization have added to the complex nature of the environmental and equity problems that challenge the development of sustainable programs and policies. This is especially true when considering the political and social decisions during the age of industrial development and European urban settlement.

These values include the use and potential exploitation of natural resources in pursuit of profit, with profit deemed a proxy for a social good. Profit as a motive is a powerful force and has arguably provided a level of goods and services that has extended the quality and length of human life. An economy based on the idea that profit is a proxy for an unquestioned social good, however, can become an economy operating in contradiction to sustainable approaches and practices. For example, critics point out how these values ignore the creation of waste and pollution, and their inevitable environmental consequences on the planet's ecosystem. Examples include the dumping of treated sewage into rivers and oceans, the creation of radioactive waste as a by-product of nuclear energy generation, and the release of the greenhouse gas carbon dioxide along with the carcinogen dioxin because of the incineration of certain petrochemical products.

In addition, the economies of most industrialized nations, including the United States, are driven by consumerism that is deemed necessary to drive an economy based on maximum use of natural resources for as much profit as possible. The policies, customs, and personal practices founded on industrial values have led to a rapid diminution of natural resources, geometric increases in pollution problems, and near gridlock in the ability of contemporary policymakers to address these policy-driven, postindustrial environmental problems. With rapidly rising human populations and more pressure to consume natural resources for economic development, the overall damage to the planet's ecosystem could have devastating consequences, especially for the world's poorest populations.

The values that lay the foundation for sustainability are explicit concern for future generations of life in all its diversity and questioning consumerism. It is no longer possible to exploit natural resources relentlessly and recklessly as in the past because now all life depends on them. Industrial values and the policies they formed led to our current postindustrial environmental gridlock, especially in matters regarding sustainability and equity. Because many Western industrialized societies are in decisional gridlock on environmental issues, and the environmental degradation threatens ecosystems, more people have become aware of the values that underlie industrialization and uncontrolled profit taking from natural resources. More people share the values of sustainability as issues of the environment become personal.

Human Habitation and Its Impact on the Environment

The United Nations and its various agencies began studying our destructive impact on the Earth in the early 1990s. The concept of global warming alarmed some scientists and was repudiated by others. Many governments that focused on economic development, free markets, capitalism, and private property also hotly contested the concept. The elected leadership of these nations refused to develop any policy that would decrease the transactional efficiency and corporate profits of the status quo at the time. Powerful nations, such as the United States, rejected international agreements based on global warming concepts. They rejected alternative energy development and engaged in wars to ensure an adequate supply of energy from nonrenewable energy sources such as petrochemicals. Because of the limited scale of human scientific study of the planet, it was hard to tell cause and effect under the limited conditions of scientific inquiry of the time. As global warming became observable and as the effect of humans on global warming became more accepted, the next stage of the debate was on whether global warming caused changes in climate. Even at this time, many powerful nations such as the United States refuse to develop public policies that would mitigate their impacts on global warming. Some scientists pursued the idea that global warming could be economically beneficial for some nations. For example, the melting ice of the Nordic nations might create more hydroelectric power. The primary concern is on exploring economic advantages of climate change, not on mitigating the impacts on ecosystems. This underscores the political context of many scientific inquiries and scientific models of causality.

At the turn of the century, the United Nations began the Millennium Project in earnest to bring together the leading scientists of the time to study ecosystems around the globe, especially those that were most at risk of irreparable damage. These scientists embrace uncertainty, transparent environmental transactions, and full cost accounting of all human impacts on the environment. They seek to understand fully the systems of nature on which all future life depends and to develop models of sustainable development. Their work is often controversial because it directly challenges long-held traditions around religion, systems of government, and beliefs about the environment.

Human impacts on the environment often come from human uses of the land, air, and water. Waste and pollution are by-products of human development. The life systems of nature produce no products that are not used by other ecosystem processes. Ecosystem processes are thus described as closed loops. As human development increases in scale, so do waste and pollution and their consequences for ecosystems on which all life depends. The challenge of sustainable human development is whether it is possible to develop and grow within the limits of our ecosystems. Is it possible to meet current needs by using ecosystem services without destroying the potential of these systems to supply the needs of future generations?

Human settlements first began to grow near confluences of fertile soil and fresh, navigable water. Water fosters agriculture, and it is a transportation route linking communities together. As communities grow, so do discharges of waster and run-off surface water from developed surfaces. These carry concentrations of chemicals, bacteria, and other human waste products into surrounding ecosystems where they can and do cause unwelcome consequences for life systems and living beings within them. The tendency of human settlements to grow causes water pollution of surface and groundwater, loss of soil, and salinization of soil; threatens surrounding farmland; and disrupts species habitat. This dynamic is accelerated by the use of fossil fuels to provide fundamental energy to run machinery of every type. Population and technological growth caused major ecological imbalances with many human habitations, later called cities, towns, and villages.

Humans have manipulated ecosystems so adeptly that we now threaten their health. The evidence of widespread human-driven destruction of ecosystems is not in doubt. The scientific evidence from the world's scientists is in. They have spoken clearly about the measurable damage from human activity, including climate change and an increase in poverty. There is uncertainty about the speed that collapse will take and its full consequences. Human sentience may allow us moderation and management rather than exploitation of the environmental condition. As we become more aware of our situation and the condition of the earth systems we depend on, we have the capacity to choose to act together to avert disaster. The capacity to see things through the lens of a shared, fragile system may be the most effective way to stimulate change across levels of behavior.

Evidence of the toll human activity has taken on nature's life systems is sobering. Human growth and development have had a significant environmental footprint, and the projected direction of growth raises concerns about whether future generations will be able to meet their fundamental needs from the life systems of our planet. The Millennium Ecosystem Assessment describes the consequences of ecosystem change for human well-being. It describes the condition and trends in the world's ecosystems and the services they provide. Its four main conclusions are:

1. Human demands have caused a huge loss of life and diversity on Earth.

2. Some human communities have benefited tremendously by the changes made to ecosystems; others have experienced increased poverty that will affect future generations.

3. The degradation of ecosystems and the increase of poverty challenge our abilities to meet goals for eliminating poverty and restoring our environment.

4. Reversing the degradation of ecosystems while meeting increasing demands for their services will require substantial changes.

The ways we conduct business, the scale of these processes, and the way we structure our economic systems are significant drivers of these consequences. For example, in the United States, 45 percent of all carbon dioxide emissions from fossil fuels come from commercial and residential sources, 21 percent from residential sources, and 33 percent from transportation (including both commercial and personal uses).

Protection of our ecosystems, and the services that they provide, requires changes in the way we do business and the foundations of our economic systems. This is equally true in both developed and developing nations. These changes are most effective at the level of our economic values—our ways of conceptualizing growth, prosperity, and happiness. In the context of the Millennium Assessment conclusions, businesses and economic systems with the greatest impact on ecosystems must become leaders in restoring those ecosystems whether these are located in the United States, China, or India. Their futures depend on doing so, as well as the stability and security of all that depend on those ecosystems for their lives and livelihoods. In a real sense, we are so interconnected now that no ecosystem can be allowed to become a sacrifice zone. The duty to care for ecosystems extends to all of us, and all of our future generations. With population growth projected to rapidly increase, even ecosystems in currently unpopulated areas need to be monitored and protected. This is another hallmark of sustainability.

What is driving us beyond the limits of our ecosystems and their ability to support human life? The answer is to be found in how we contemporary humans choose to make our livelihoods, what we consume, and how many of us do these things. These are the fundamental questions of economics and business, consumption, and population. This volume explores sustainability in the context of economics, business, consumerism, and population. When we use up the resources provided by our ecosystems faster than they can be replaced, if they are replaceable at all, we deplete the stock of what is available to run our economies.

Economies differ widely from place to place based on many factors including local geography, labor, markets, transportation, taxes, and subsidies. Our economies are made up of large and small participants. Some participate in our economy by working and consuming; others participate by providing goods and services. Some participate to earn profits, and others contribute to our economies without profit motives.

Our present economies have changed our planet's climate and altered the systems that support all life. The energy that fuels our economies depends on combustion of fossil fuels that accelerate the phenomenon of global warming. Globalization of our markets and communications has contributed to an increase in the production and consumption of goods that are exorbitant consumers of resources. The consequences of these changes have hurt the world's vulnerable people and communities the most. These two trends continue to worsen rapidly, and on a global scale. In our contemporary world, economies are connected by many links including global communications networks, transportation networks, and banking and finance networks, creating ever-larger global marketplaces in which goods and services are traded. This trade prompts ever-growing demands on our ecosystems that provide the resources and energy necessary to provide for our human enterprises and communities.

Sustainability in the context of economics and business recognizes that all livelihoods have their foundations in the systems that support life on earth. Sustainable economics and business also recognize that our economies have compromised the ability of these systems to provide for future generations because businesses and governments undervalue or completely ignore these essential systems. The constellation of challenges of sustainability in this context is to find new ways to make livelihoods, to eliminate waste and pollution in what we make and consume, and perhaps most significantly, to change our thinking about our needs, wants, and desires—in short, what it is that makes for human happiness. One of the most powerful ways of doing this is to change our business planning cycles from short-term periods accounting for profits, but not costs, to longer term visions of accountability for resources to future generations. Some express this as changing from exclusively profit measures of success to include community and environmental accountability as well. Whether couched in terms of reverence, faith, and ethics or accounting terms, sustainable economics and business accept a duty to future generations to change and restrain our choices in ways that value the systems on which all life on Earth ultimately depends.

At the theoretical levels, some of the most transformative economic ideas include accounting for the true and full benefits of ecosystems, and requiring accountability for the true and full costs of our economic activities. This is one of the cornerstones of the United Nations Green Economy Initiative; The Economics of Ecosystems and Biodiversity part of the project will be to assess the economic values of ecosystems and biodiversity. By valuing these systems and processes accurately, we will be able to appreciate the significance of their loss. The hope is that this will change our thinking and shift our economies toward ecologically restorative processes.

The pressure to continue our present patterns of unsustainable economics and business comes from increased poverty, even while environmental degradation escalates, leaving the poorest worse off. Another source of pressure to continue this trajectory is greed. Sustainable development strategies have confronted these sources of pressure to develop. The controversies about sustainable economics and business emerge from these conflicts.

What is the responsibility of business and economic policy for these conditions? What is their role in restoring ecosystems and healthy communities? The developing language of sustainable economics speaks to both questions: why business has a role and a responsibility to ecosystem and community health. This language also is developing concepts, methods, and stimulating technology essential to doing business without degrading ecosystems. The duty to alleviate poverty is also slowly infiltrating the discourse of sustainable business.

Reference

2009 Greenhouse Gas Inventory Report (1990–2007). US EPA, April 15, 2009.

THE ROLE AND RESPONSIBILITY OF BUSINESS AND ECONOMICS FOR SUSTAINABILITY

Economic activities conducted by individual businesses and industries allow human life to become more secure, pleasant, and fulfilling. We humans have adapted our activities in order to grow. Yet this very success contributes to critically degrading ecosystems, together with growing populations and increasing demands on finite resources. Agriculture has allowed humans to do more than merely survive; secure production of food surplus allows us as a species to do more than survive and replace our populations. The impacts of agribusiness on ecosystems, however, are alarming. Fertilizers, pesticides, and other chemicals contribute to a dangerous slurry of runoff that can pollute our fresh water resources. Livestock factories created lagoons of waste that threaten surrounding human habitation with waterborne disease. This is the source of much of the tension and dissonance between the three dimensions of sustainability. Pursuit of short-term maximization of financial profits may lead to degradation of shared ecosystems like water and air. Degraded ecosystems harm lives dependent on them, especially overburdened and vulnerable populations such as communities of color, children, and the elderly and endangered species of animals and plants.

Sustainable development refers to growth in the human quality of life without irreparably damaging ecosystems that future life will depend on. In most Western nations, growth often refers to economic growth. This means that the population's per capita and disposable incomes increase. It is usually limited to economic growth for the present generation and does not fully consider economic growth in the future. It therefore generally fails to consider the effects of rapid population growth and the context of future goods and services. Traditional economic growth also fails to consider the systems of nature that constrain economic growth as populations increase and consume more natural resources. These natural resources can run out, become depleted, and fail to replace themselves. Systems of life on which future generations depend become irreversibly impaired and therefore traditional patterns of economic growth are not sustainable.

This is especially true if both population and consumption of natural resources grow simultaneously. A downward spiral of environmental degradation and lowering of economic standards can begin. If consumption becomes a measure of social status, the impact on the environment can be quickly seen. If the economic processes are inefficient and create waste that erodes environmental integrity, then quality of life begins to erode.

Waste or Resources?

Waste of all types increases with both population and economic inefficiency. Waste from one source can be a good or commodity to another source. In terms of production processes, reusing what is waste to develop another commodity is called closing the loops of production to eliminate waste through reuse and increasing economic efficiency. *Waste* itself is a term of art that is carefully analyzed in the process of sustainable development. Wastes can accumulate and overload ecosystems in ways that overwhelm it so much that it threatens to irreparably harm the systems of nature on which future life depends. For example, it takes tens of thousands of years for nuclear wastes from energy production to safely reenter the environment without causing irreparable damage. Ultimately, ecosystem integrity requires a balance of human needs and impacts with systems of nature over the time scales of nature. Certain ecosystem services, like fresh water, wood, and clean air, can become depleted while solutions are lost in controversies about religion, capitalism, private property, and endangered species.

Three Definitions of Sustainability

Sustainability as Economic Growth

One version of sustainability considers mainly economic growth. This variant of sustainability calls for maximum economic growth, and the environment is a necessary precondition to this goal. This version of sustainability has dominated thinking in the United States more than other regions and areas of the globe. Under this version, indicators of success are found in measures like gross national product or gross domestic income. These are measurements of the dollar value of certain products and transactions. These types of financial indicators are not qualified by the costs and externalities they may impose. They do not distinguish between products and services that are related to harms that should be avoided such as cleaning up oil spills and brownfields, and those that are related to increasing the quality of life such as educational spending and preventive health care. These distinctions are important to some societies as indicators of whether progress toward sustainability is actually being attained. This has led to the creation of different measures of success such as gross national happiness, which considers the financial gains in the context of other indicators of human happiness and well-being (see Terms and Definitions in chapter 2 of this volume).

This version of sustainability may seem at odds with the stated agenda of many modern-day environmentalists. It is at odds with most principles of equity because maximum economic growth for the present growing population will consume irreplaceable natural resources. Equity also asks the question of economic growth for whom. What is the environmental decision-making process by which irreplaceable natural resources are developed profitably?

Sustainability as Growth within Limits: Ecology and Equity

A second version of sustainability is again placed in the context of economic growth, although in this context it does not seek to maximize economic growth. Economic actions that are clearly detrimental to the environment are avoided. Instead, a set of goals including security and peace, economic development, social progress, and good government is advocated. This model of development has influenced global thinking about development more than thinking within the United States. The significant contribution of sustainable development to this model is that it clearly defines all goals in terms of the limitations of our ecosystem health.

The most significant challenges of implementing this approach to sustainability include poverty, precaution, and risk assessment. Specifically, this approach may founder on questions of how to respond to issues of fairness and equity, especially when actions will harm poor countries and poor people more than it will harm others. It may also go wrong if it fails to require precautionary measures, especially when scientific proof of causality is absent. It may also fail based on its approach to the assessment, perception, and management of risks.

The problems of poverty compound issues of sustainability because changes in energy, technology, or products often raise the costs of these items. Business and economists may not recognize or account for these consequences based on calculations of profitability, but they should because harms inflicted on poor people and poor countries result in environmentally unsustainable crisis decisions. For example, climate change requires that businesses and industry move away from fossil fuels. Business and industry account for at least two-thirds of greenhouse gas generation. So far, however, greener energy costs more because it will require investment in infrastructure. If the costs of converting to greener energy sources are passed on to individuals and households, poor and limited income individuals and families will suffer. In some areas this will mean increased use of fuels like wood, which puts more smoke and particulate matter into the ambient air in our communities. When these conditions persist, they cause health problems for elderly people in the form of heart attacks, strokes, and illness, as well as increased episodes of asthma-related emergencies. These conditions will eventually affect business and industry in the area, too, as they lose employee productivity and pay increased costs for health-related benefits.

Another argument that business and industry should consider the consequences of their acts on poor people and communities was made by Henry Ford. Ford urged that all of his employees should be able to afford the automobiles that they manufactured, so that they could buy their own products, ensuring their continued economic viability.

This approach can also fail at the level of implementing the precautionary principle. This principle requires that reasonable measures be taken to protect against potential harm to the environment, even when science is uncertain as to causality. This requirement changes the way many economic and business decisions are made because it requires that the proponent of action such as a business take steps to prevent harm before it occurs, rather than waiting for harm to occur and then compensating victims for it. In addition, this approach shifts the burden of proving harm from individual victims to the business that wants to act. Again, businesses and industries may not want to recognize or be responsible for these consequences in advance of action, but they should do so. By making environmental consequences part of their planning processes, better economic decisions may be made, internalizing true and full costs of conduct rather than encountering massive costs from environmental disasters later, and perhaps shifting these costs onto innocent and vulnerable people and ecologies.

Finally, the success of this approach to sustainability depends at many stages on a complete and accurate assessment of risks and consequences of proposed actions, and the appropriate management of those risks and consequences. Risk assessment and management in turn require inclusive thinking and participation by those affected by decisions. This kind of construction of risk is not typically the way businesses or risk managers approach decision making, but they should. There are new forums and formats for making this happen. By thinking broadly about the risks and consequences of action, business makes better decisions. Including community as a partner in those plans also ensures healthy, mutually supportive relationships over time. Good neighbor agreements, community benefits agreements, and other types of positive community interfaces with business are becoming part of the way sustainable businesses stay sustainable. ***See also* Volume 3, Chapter 5: Education on the Environment for Sustainability.**

SUSTAINABILITY AS ECOSYSTEM DEPENDENT

A third version of sustainability is rooted in the impacts on the environment, generally including humans. In this perspective, ecosystem impacts subordinate all other considerations including human equity and economic survival. It is tied to human impacts on the land, air, and water. It requires an understanding of all our environmental impacts, including the cumulative impacts of everyday activities. This version of sustainability explicitly and implicitly challenges the existing value foundations of existing social, economic, and political institutions. Many environmentalists feel that this is the one true version of sustainability. They argue that because all life is dependent on the conditions of the ecosystems that support life, economic and social concerns are justifiably secondary to ecosystem health. This relationship to sustainability is captured in the image of nested baskets. The environment is the largest, most comprehensive basket. Within it, all human activity is located, including human economic enterprises and human communities. Economic enterprises are represented by a basket nested within our environment and its webs of life. Human individuals and their communities rest within both these two baskets, relying on each for their livelihood and support.

Frank Lloyd Wright and Usonian Houses

Frank Lloyd Wright is best known as a renowned architect who constructed exquisite private homes that seamlessly integrated elements of nature into his built environments. His philosophical vision embraced much wider projects including diminished use of construction materials and affordable housing for low- and middle-income people. He used the term *usonian* to describe his designs for this type of alternative housing. Usonian stands for United State of North America, reflecting his belief that the term *American* inaccurately tends to exclude Canada and Mexico.

Law Office Sustainability Tools

Service-oriented businesses may not recognize how they affect the natural world. Operating at a distance from the land and from manufacturing, such organizations may choose to assume that they are not implicated in damage done to the environment. The first step for such organizations is awareness of their environmental footprint, and the resolve to diminish it. Law offices have developed tools to assist the profession in developing a model of sustainable service sector operations. These tools and practices focus on purchasing decisions, waste reduction and recycling, energy usage, reducing the carbon footprint of travel and commuting, tenant improvements, and a commitment to continued environmental and sustainability education.

ADAPTIVE ECONOMIC SYSTEMS: THE RESTORATIVE ECONOMY

The emerging language of sustainability in business and economics terms conceptualizes these changes by mirroring the designs and dynamics of nature, sometimes-called biomimicry. For example, technological advancement in robotic mobility is made by copying nature. This rests on a simple observation that in nature there is no waste or pollution because the product of any individual process is a useful resource for some other life cycle. These are often described as closed loops. When business processes eliminate waste and pollution, they have made a significant step toward sustainability.

In Chapter 2 we present a section on definitions of some of the emerging terms. In the area of sustainable development and ecosystem exploration, the current language often falls short of an adequate explanation. Differences in culture, education, values, and experiences with nature all create difficulty in communicating some of these concepts. This is one of the difficulties with defining sustainability. It can be overused in present economic values to market an item without the related concepts of the true measure of environmental impacts through the lifecycle of the product. The abuse of the term *sustainability* does detract slightly from its meaning, but the early definitions still apply and cast manipulated definitions in the light of so-called greenwashing.

Gro Harlem Brundtland

Gro Harlem was born in Oslo, Norway, in 1939. She received her medical degree from the University of Oslo and earned a master's degree in public health from Harvard University. Although her focus was on women and children's health, she became interested and involved in issues of poverty, population growth, food security, and public health on a global scale. She was a well-known doctor of rehabilitation and a prominent member of the Norwegian Labour Party. In 1974, after working for the Department of Hygiene and Oslo Municipal Board of Health, she accepted a position as minister of the environment for Norway and in 1983 was the first woman elected prime minister of Norway. While serving in this position, the United Nations secretary general asked her to create and head the World Commission on Environment and Development. From this post, she wrote "Our Common Future," also known as the Brundtland Report, which is often cited as the foundational document for sustainable development.

The report emphasized several key concepts, including the interconnectivity of the world and the need for increased international multilateralism and cooperation to solve root problems. The report integrated environment and development, "[T]he environment is where we live, and development is what we do in attempting to improve our lot within that abode."

Reference

World Commission on Environment and Development. 1987. *Our Common Future: Report of the World Commission on Environment and Development*. New York: Oxford University Press (available on line at www.un-documents.net/wced-ocf.htm).

Reference

Benyus, Janine M. 2002. *Biomimicry: Innovation Inspired by Nature*. New York: Perennial.

Parr, Adrian. 2009. *Hijacking Sustainability*. Cambridge, MA. MIT Press.

Speth, James Gustave. 2008. *The Bridge at the Edge of the World: Capitalism, the Environment, and Crossing from Crisis to Sustainability*. New Haven, CT: Yale University Press.

ECONOMIES AND THE WEBS OF LIFE

Our economies are the human activities we engage in to support our needs and desires for more security and comfort. Economies are composed of all the various things we make and consume for profit or for exchange. Participants in the economy include all of us in many different roles, from consumers to workers, to eaters, to owners and managers. Participants also include organizations of different sizes and missions, using different resources and producing all manner of items. From this broad, expansive definition, it is clear that our ways of earning a living from the earth are rapidly depleting the very systems we all depend on. Businesses that use resources or leave pollution in ways that diminish the health and well-being of future generations should consider whether they are conducting business in an ethical or fair way (see Volume 3). Climate change driven by human emissions will eliminate many plants and animals from our world forever. In developed countries, current and future children face life-shortening epidemics of obesity caused in part by food and product additives like transfats, and some forms of plastic packaging, as well as unsafe exterior conditions to play in. These conditions are related in substantial part to the way we do business, the resources we use, and the pollution and poisons we use to make and distribute products.

Reaping the benefits of those methods of production so intensively now and leaving behind ecological loss is an issue of intergenerational injustice. Sustainability unequivocally demands justice for future generations. Many businesses have not looked at their economic activities in this way. They may have recognized certain economic costs and risks to their current practices, but not to a future generation. That kind of limitation in accountability and vision leads businesses and whole economies to shift major, irreparable harms onto the backs of those future generations and contemporary generations who they do not recognize. These externalized forces fall on those who are powerless to resist the public heal effects of ecological degradation such as poor people and poor communities.

THE USE OF RESOURCES AND ENERGY

The majority of the materials we consume go into the construction of buildings. The largest part of the energy we consume is consumed within built structures. The majority of our building products and energy are based on fossil fuels. This makes our current economies and infrastructure dependent on oil, gas, and coal. Human burning of these fuels is contributing to climate change, loss of biodiversity, severe weather incidents, and human health consequences.

The largest part of the waste that goes into landfills is from construction. The largest part of all the energy we consume goes into buildings and industry, around 74 percent combined. When businesses reduce material and energy usage, they can have an enormous impact on the use of virgin and nonrenewable resources. Industry (including agriculture) is the largest consumer of resources and energy. The challenge of sustainability for businesses is to lessen their ecological footprint and perhaps beyond that to restore ecosystems. Globalization of markets, multinational corporations, wars, and consumerism have increased the environmental footprint of industrial use of our finite ecological systems; and the trajectory of use, especially in light of population increases, is pointing toward collapse. The weakness of governmental power to slow or change this trajectory or footprint is apparent in ecological outcomes (see Volume 1). Some industries have chosen to voluntarily embrace sustainable business practices like sustainable production to the extent that they increase long-term profits. Completely new branches of design and engineering have developed based on the conviction that we must remove waste, toxins, and pollution from our stream of production.

Businesses can achieve these results in more than one way. Two ways to reduce energy and materials usage are, first, to design systems that do not use materials unsustainably and do not use energy unsustainably, and second, to create consumer demand for products produced this way. The profit motive must not subordinate environmental protection. The United Nations has promoted sustainable production and consumption methods as part of its Environment Program (see UN Environmental Programme in this volume). Many people and governments have developed multiple strategies to achieve these goals. For example, Dr. Karl-Henrik Robèrt developed a framework for thinking about sustainability and human activities based on the laws of thermodynamics. He called his method, "The Natural Step," and many businesses have voluntarily chosen to reorganize their operations and products with this framework for sustainability in mind.

The Natural Step is an international nonprofit organization based in Stockholm, Sweden, with national branches in many countries around the world. The Natural Step philosophy teaches that in order to become and remain sustainable, a society must maintain four fundamental conditions known as "System Conditions":

1. We must maintain the concentrations of substances extracted from the Earth's crust.
2. Concentrations of substances produced by society should not be produced faster than they can be broken down and redeposited in the Earth's crust.

3. Degradation by natural physical means must be maintained.
4. Conditions must be maintained that allow humans to meet their basic needs, and these conditions must be fair and equitable.

From The Natural Step Canada www.naturalstep.

To help complex organizations reorganize themselves using this philosophy, the Natural Step organization has developed a process called Backcasting or the A-B-C-D Analytical Approach. Backcasting involves translating each of the four system conditions into outcomes specific to the organization. Then, the organization is asked to link present conditions to those outcomes in a series of steps or goals in order to achieve sustainability within that organizational system.

Consumers may influence materials, toxics, and energy usage when they create a demand for sustainably produced products and services. When consumers are numerically significant enough or when they are independently important, their preferences cause changes in products and production processes. This is called "making a market." For example, when government or even a multinational corporation decides that

The Natural Step's Four System Conditions

1. We must maintain the concentrations of substances extracted from the Earth's crust.

2. Concentrations of substances produced by society shouldn't be produced faster than they can be broken down and redeposited in the Earth's Crust.

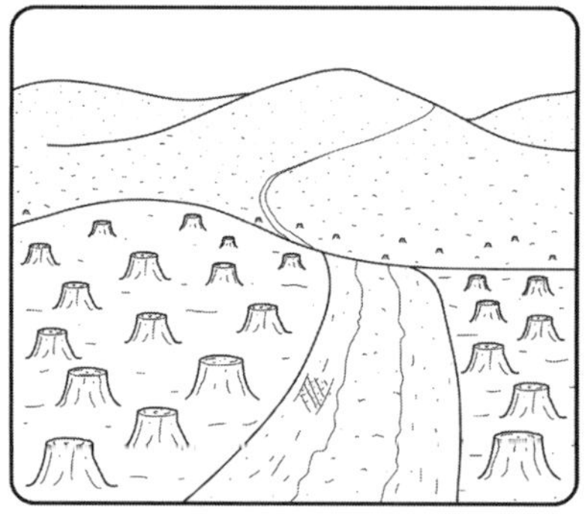

3. Degradation by physical means must be maintained.

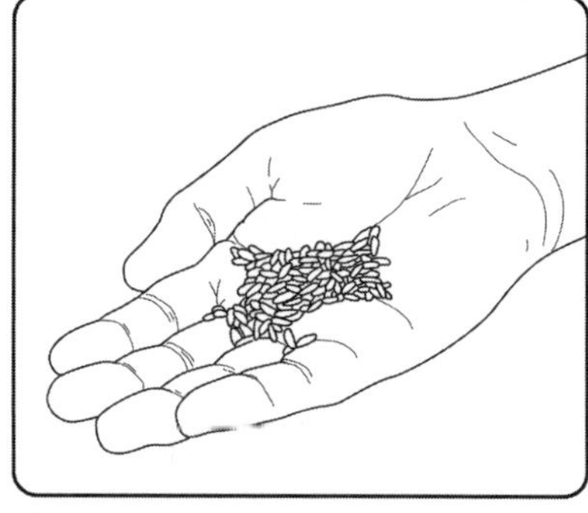

4. Conditions must be maintained that allow humans to meet their basic needs, and these conditions must be fair and equitable.

FIGURE 2.5 • Four systems conditions of sustainability. Illustrator: Jeff Dixon.

The Natural Step Leadership

Dr. Karl-Henrik Robèrt is one of Sweden's most successful cancer scientists. He has written books about leukemia, lymphoma, lung cancer, and cancer's clinical implications. His research looks at how human cells are damaged. It has led him to initiate an environmental movement called The Natural Step (TNS). TNS is based on a set of principles called the systems conditions and calls for systems thinking and broad-based solutions.

When Robèrt and a group of scientists and political and business decision makers in Sweden finished compiling the framework for TNS, Robèrt created education materials and provided them to every school and household in Sweden. He wanted to spread the idea of system conditions and system thinking as fast as possible. Business executives from major Swedish companies such as, IKEA, Electrolux, Swedish McDonalds, construction companies, the leading supermarket chains of Sweden, insurance companies, banks, and a large number of other business corporations and municipalities have incorporated the TNS framework into their business practices.

TNS is founded on the concept that we humans are a part of nature. Robèrt makes the point that our current practices, such as waste creation, toxin creation, unfair allocation of resources, pollution, and environmental degradation, will lead to our own destruction, as well as to the destruction of nature. Robèrt criticizes the practice of needing certainty within the scientific community before we decide to act against something. For example, scientists are asked to tell the leaders what amount of mercury in a human is permissible before it becomes harmful. Robèrt argues that we design harmful products out entirely instead of trying to decide what an allowable amount is. Systems thinkers would also not consider the current elimination of mercury a solution because they view the entire system as a closed loop. Nothing is truly eliminated from the ecosystem from this systems perspective. Even if we stop using mercury today, we must ask "Where is the mercury now?" It is in filters and other products. We have reduced emissions from the chimneys, but that is not the solution.

FIGURE 2.6 • Karl-Henrik Robèrt is a cancer scientist and Professor of Resource Theory at the University of Gothenburg in Sweden. Having initiated The Natural Step movement in 1989, he was awarded the Green Cross Award for International leadership in 1999, and the Blue Planet Prize (the 'environment Nobel') in 2000. [Courtesy of the Natural Step International]

it will only use certain sustainably designed or produced products, contractors and suppliers will produce those types of products rather than traditional ones. This shift has the added effect of making sustainable products more widely available to others. Having many different strategies, an encyclopedia of sustainability engages many different businesses in a diverse economy in efforts to become more sustainable individually and build a more sustainable economy overall. ***See also* Volume 2, Chapter 4: Energy; Volume 2, Chapter 4: Energy Sources.**

References

Dauvergne, Peter. 2008. *The Shadows of Consumption: Consequences for the Global Environment.* Cambridge, MA: MIT Press.

Josephson, Paul. 2004. *Resources under Regimes: Technology, Environment, and the State.* Cambridge, MA: Harvard University Press.

Pirages, Dennis, and Ken Cousins. 2005. *From Resource Scarcity to Ecological Security: Exploring New Limits to Growth.* Cambridge, MA: MIT Press.

THE ROLE OF BUSINESS, CONSUMERS, AND INVESTORS IN ACHIEVING SUSTAINABILITY

Business plays an essential role in achieving sustainability by its impacts on our ecologies and human communities. The largest businesses, including multinational corporations, have larger economies than many countries. This scale makes their policies and practices as important as those of many governments.

Consumers and purchasers in turn shape what businesses produce. Again, size matters because the larger the purchaser or consumer, the greater the influence on the goods businesses produce, and how they produce it. This is sometimes referred to as the power to make a market in a particular type of good. For example if Wal-Mart or the U.S. Department of Defense decides to purchase LED lights or other low-energy products, its purchasing power is so significant that manufacturers will bring products to it if they are not already available. Technology can be market driven and used to decrease environmental impacts.

Finally, publicly traded corporations are financed by groups of investors. Shareholders invest their money into a corporate form because their losses are limited to the value of their shares, and nothing else. They are not personally liable for the debts of the corporation. Some investors are large groups of individuals or other organizations called institutional investors. Other investors are individual people. Corporations want to attract investors to their firms if they are publicly traded so that they will have more money to work with. In this way, investor values and principles may shape the way a corporation uses resources and accounts for its activities.

Many types of enterprises go into the making of an economy, from multinational corporate entities to nonprofit organizations. The proprietary activities of nations, states and municipalities, indigenous nation

organizations, nongovernmental organizations, and civil societies of all types contribute to the supply and demand of an economy. Many of these enterprises are increasingly involved in trying to achieve sustainability at some level. Businesses and these other institutions are responsible for the great proportion of environmental impacts. They are also trying to find ways to mitigate and eliminate those impacts so that future generations may have the same opportunities from the environment.

Corporate Social Responsibility

Corporations are one particular form of business. They are made up of investors called shareholders who put money into the corporation in return for a share of its profits, when profits are made. They do not make business decisions for the corporation. Business decisions and business policies are determined by a governing board elected by shareholders. That governing board is usually called a board of directors. The board has the power to make fundamental business decisions that are then implemented by the corporation's officers and employees.

In recent years, many forces both within the world's corporations and outside those corporations have urged these forms of business to embrace responsibilities to the environment and the communities they affect. These responsibilities often go well beyond the simple obligation to make profits for their shareholders.

References

Graedel, Thomas E. et al. 2005. *Greening the Industrial Facility: Perspectives, Approaches, and Tools.* New York: Springer Press.

Parr, Adrian, 2009. *Hijacking Sustainability.* Cambridge, MA. MIT Press.

Shareholder Initiatives

A corporation is a licensed form of business that allows investors to limit their liability for corporate activities to the value of their investment or shares in the company. Other forms of doing business such as partnerships or sole proprietorships leave their investors fully liable for company activities. This unlimited liability extends to the value of all personal assets. This form of limited liability directed an enormous amount of capital for investment to corporations. Corporate shareholders have some say in what the company does by electing members of a board of directors who have the power to determine corporate policy choices. Boards of directors may choose what a company does, and how it does business.

The ability to elect directors is accomplished by allowing shareholders to vote during regularly held meetings. Some activists have advocated investment in corporations as a way to elect leadership for sustainable choices within corporations.

Reference

Lewis, Sanford J. 1993. *The Good Neighbor Handbook: A Community-Based Strategy for Sustainable Industry.* Waverly, MA: Good Neighbor Project.

Nonprofit Organizations

Nonprofit organizations often engage in many important economic activities, but they do not pursue their activities for a financial profit. Nonprofit organizations include churches, schools, environmental groups, and many similar organizations. These organizations participate in the economy by pursuing their missions while paying for their costs of production and labor. These organizations often do not have to pay taxes. They can be powerful partners with government to achieve socially and environmentally beneficial goals such as education, site cleanup, and community capacity building. They are also called "civil society organizations." *See also* **Volume 3, Chapters 2 and 3**.

Utilities

A utility is a company designed to manage an essential resource for the benefit of an entire community. Some utilities are operated by local governments, others can be owned and operated by the community members they serve (called a co-op), and some may be privately owned and traded on the stock market. Not all utilities are nonprofit companies, although many are not for profit. Utilities are special kinds of businesses because they are providing essential services to communities and are not subject to competition from other businesses for the profits that they earn from these services. This monopoly over resources and services is subject to public oversight even when the utility is privately owned, because the resource is supposed to be managed consistently with the best interests of the community.

Publicly Owned Treatment Works

Creating utilities with such monopoly powers was a significant tool in the development of the infrastructure delivering power and clean water to individual households and businesses in the United States. For example, many states created utilities called publicly owned treatment works (POTW) to provide and maintain clean fresh water to their communities. POTWs are wastewater treatment plants that must remove a wide variety of pollutants and waste from freshwater sources. This is a significant success in development and public sanitation and health. Keeping waste and fresh water separate is a continuing challenge for all human development.

Power-generating utilities successfully generated power sufficient to create great economic wealth and to deliver power to most households; however, they also became responsible for creating significant waste and pollution of air and water. Several state attorneys general sued several coal-burning utilities for their contribution to acid rain that was injuring the landmarks and ecologies of their states. The contribution of energy utilities to climate change is affecting water resources through the processes of rising seawaters, salinization, droughts, sedimentation, turbidity, eutrophication, and increasing contamination in freshwater

sources. In response to these problems, public and private utilities have been required to convert a percentage of their operations support and develop nonpolluting energy resources such as wind or geothermal power.

In addition, many states are exploring ways to use the business form of a utility to develop the new infrastructures needed for production and distribution of new, clean energy sources such as solar power. (For more information, visit www.solarelectricpower.org to view solar power utility programs by state.)

Academia and Sustainability: Creating a Constituency for Sustainability

There can be no lasting change in a society without an educational policy and program to support that change. This was the operating theory that has guided many developing countries struggling to develop in a postcolonial, postindustrial world. The great economies of China and India developed after colonialism and the growth of industrialism relied on their abilities to educate their population to perform complex, nonmenial work.

The task facing developed countries in the struggle to achieve global sustainability is to do the same at a time when commitment to public education financed by local property taxes has weakened. In developed democracies, education has not kept pace with the public need and interest in sustainability as an area of study and coherent policy development, even as the support for public education has dwindled.

References

Association for the Advancement of Sustainability in Higher Education, www.aashe.org/index.php.

Corcoran, Peter, Wals Blaze, and E. J. Arfen. 2007. *Higher Education and the Challenge of Sustainability: Problematics, Promise, and Practice.* New York: Springer.

Business and Equity

Values have an economic and moral meaning. Both economic and moral values are reflected in how a society organizes its economic concerns and distributes its resources. Definitive values tie together the entire concept of sustainability. First, there is an explicit concern for future generations. The major documents on sustainability all refer to a principle of resource use that protects the interests of future generations of humans and nonhuman species as well. The central defining characteristic of sustainability is its concern that contemporary humans conduct themselves in a way that protects the interests of future generations of humans: a specific kind of intergenerational equity. In this way, intergenerational equity is the central, core value of sustainability. This is reflected in a variety of ways including the concept of carrying capacity, sustainable development, and resource management. ***See also* Volume 2, Chapter 2: Language of Sustainable Business.**

THE NEED FOR ECONOMIC DEVELOPMENT: POVERTY

If businesses linked into a global economy are not producing in a sustainable way, globalization threatens to intensify pressures on ecosystems, resources, and energy in ways that spiral ever more quickly toward ecosystem collapses. Meanwhile, the social and environmental dangers of globalization may be masked by the phenomenon of externalization of costs until the harms are disastrous. The ability to harness the power of markets to develop without triggering collapse of critical ecosystem functions is another area of intense controversy, even though increasing costs of transportation of goods may undercut the salience of global markets.

Some people have questioned the need for further development at all. The current challenges to our sustainability are the result of how we have chosen to develop. Further development, some may say, can only make matters worse. In countries ravaged by poverty, hunger, unsanitary conditions, and inability to shelter their people, however, development is essential to ensure basic human rights. Some people question the wisdom of having more children born in countries without basic means of life. Birth rates are not population rates, and population rates grow as a function of immigration as well as live births. Similarly, population rates fall as a function of mortality rates and emigration. HIV, wars, and environmental disasters have caused dramatic depopulation in some already poor nations. What is unclear is the impact that population alone has on the use of resources. What is clear is that population alone does not tell us how intensely a culture or society uses, wastes, or poisons its ecosystems.

Global leadership requires the recognition of the different contexts of sustainability, poverty, history, and ecology. Sustainable development strategies mean different things under different conditions. In developed areas, the challenges may mean slowing down growth and reducing waste, pollution, and consumption. In developing areas, it may mean developing or renewing areas and developing new production using sustainable methods and products. The idea of differentiated leadership roles tries to reconcile these differences without hypocrisy. For example, the United States uses far more of the world's resources per capita than any other nation on earth. Leadership for sustainable development in the United States must reduce consumption and waste. Countries like Haiti consume far less per capita, although they are much more densely populated. Leadership for sustainability must develop new sources of food, energy, clothing, and basic shelter in ways that maintain and restore this island's decimated ecosystems. ***See also* Volume 2, Chapter 4: Poverty; Volume 3, Chapter 1.**

CHAPTER 2

Definitions and Contexts

The language of sustainability occurs in different contexts. Words used when describing aspects of sustainability give it meaning when defined in a given context. The definitions and explanations that follow give meaning to the language of sustainability in the context of business and economics.

LANGUAGE OF SUSTAINABLE BUSINESS

Businesses Types

Businesses include organizations of different sizes and missions. The U.S. government estimates that 99.7 percent of its economy is composed of small businesses. These are businesses with only a few dozen employees and owned by an individual or family. They can be organized as sole proprietorship, partnerships, or private corporations. The rest of the economy includes corporations, utilities and cooperatives, and large partnerships. Some of these are publicly financed in the sense that they are allowed to raise funds by trading their shares of stock on the open market. Most businesses are privately owned and do not have access to the public markets to raise capital for their operations. The financial significance of the few businesses that use public markets to raise capital, however, is enormous; in many cases, they are responsible for economic operations far greater than all but the largest nations.

The scale of a business operation makes a clear difference in its environmental footprint, but small businesses grouped together by environmental and economic impacts have a combined significance that is sometimes ignored because of their relatively small individual size. Sector-based approaches to issues of sustainability are able to capture issues of scale in a different way by focusing on the distinctive characteristics of a particular industrial sector.

The way that businesses are organized legally can also make an important difference in the way that decisions relating to sustainability are made. Corporate decisions must be made by boards of directors and implemented by corporate executives. Outside the corporate form, decisions about a business operation may be made and implemented by individual owners.

Standard Industrial Codes

Businesses and industries are classified by Standard Industrial Codes. Each code represents a type of business or industry. In the United States, Standard Industrial Codes were established in 1987 for use with the Occupational Health and Safety Administration, a federal agency. The role of this federal agency is to protect the health of workers on the job. Standard Industrial Codes are now being replaced by the North American Industry Classification System, which has broader international application and comparability.

The North American Industry Classification System groups businesses and industries into 20 sectors using a six-digit code for 1,170 industry classifications. The Standard Industrial Codes had 10 sectors for 1,004 classifications. The North American Classification System was developed by Mexico, Canada, and the United States under the North American Free Trade Agreement. In the United States, the Office of Management and Budget, another federal agency, assumed primary responsibility for the new groupings. Other countries, such as New Zealand and Australia, have their own industrial categories.

In terms of sustainability, these measures are important for measuring how sustainable an industry is and how sustainable an industry can be. Many industries focus on their suppliers and their environmental impacts. These codes help to set up supply chain analyses in international business transactions.

A sustainable supply chain is a system of grouped business activities that add value for all stakeholders throughout the lifecycle of the products and processes. Business models for supply chains are evolving rapidly today. Many businesses such as retail stores have many supply chains. Supply chains that are arguably sustainable help businesses improve and expand customers. A controversial issue regarding supply chains is when to "outsource" a part of the supply chain to another nation with another, often less expensive, labor source. Management of sustainable supply chains often includes codes of conduct for the suppliers, supplier questionnaires, training, and monitoring. Supplier information management is essential for sustainable supplier management. This requires the complete lifecycle of data to provide businesses with important information about supplier compliance and capacity. A major problem is that functional environmental transparency of supplier information across business type and nation is very limited.

The codes raise questions that include how far up the supply chain and how far down the product use should a particular business examine to determine its product lifecycle and its sustainability. New market values of environmental transparency and growth of environmental regulations all across the supply chain have put issues of sustainability to the forefront.

The exactitude of these industrial codes can have a direct effect on how exact the industrial sustainability analysis can become. A complicated question is how to analyze industrial sustainability in multiproduct, multisupply chain industries. In an increasing global economy with goods and services coming from places with either no industrial code classification system or ones that are not comparable, it is difficult to evaluate industrial environmental impact for a product lifecycle. Generally, lifecycle analysis takes a long view of the entire product processes and the full environmental impact. Typically, this starts with the extraction of the natural resources used in the product, any refining of these resources, how these refined resources are processed for various uses, how the products are actually used, and then how they are reused, recycled, or put in the waste stream. Industry codes can help begin a process of full cost accounting throughout the lifecycle of a product.

On a domestic level, these industrial and business classifications help tailor environmental regulations and voluntary environmental agreements to a specific industry. Basic environmental regulation often has a one-size-fits-all permitting system where all industries get the same basic permit for air or water emissions. When environmental regulations become more sophisticated, they tailor their rules to fit the supply chain, industrial processes, and products of a specific industry. Because trade associations are often industry specific, accurate industrial classifications also allow them to have a greater role.

Many sustainability advocates consider the development of accurate and comparable industrial codes to be very important in sustainable development.

Biomimicry and Ecosymbiosis

Natural cycles use the products of other cycles to create new cycles. In this way, there is no waste, no pollution, and no exhaustions of resources in a natural cycle. Nature's production cycles are described as closed loops. Biomimicry means copying natural process or making products inspired by these processes. Businesses are currently using biomimicry to make honeycomb-based glass windows, solar cells charged by photosynthesis, and self-cleaning paint based on lotus leaves and their hydrophobic characteristics.

The fact that human communities and enterprises often create waste and pollution as a product of their activities is the basis for the assertion that human life is unsustainable. The elimination of waste and pollution through design that imitates nature is at the core of economic thought and business plans that use biomimicry to achieve increased levels of sustainability. Waste and pollution are eliminated in all these designs, fundamentally through concepts that imitate nature's design pattern of closed loops in which the products of one process are used in another new process.

When businesses are grouped together or paired in cooperative relationships to use otherwise wasted by-products to provide resources for new products, this deliberately designed symbiosis re-creates closed loop biomimicry. Businesses grouped together to make mutual advantage of these factors are sometimes called ecoindustrial parks.

These design approaches can be applied to a variety of businesses beyond manufacturing, including those businesses offering services. Service-oriented sectors of the economy have a less obvious impact on resources and ecosystems. Manufacturing inputs, throughputs, and outputs are observable and quantifiable. As developed nations move increasingly toward service-oriented economies, we must consider how service sector businesses affect our environment. These impacts go far beyond recycling including purchasing policy that affects suppliers and subcontractors, landlord tenant relationships that may improve the built environment and energy consumption within it, commuting policies, and carbon-generating activities.

The core analogy making our economic systems mimic our biological systems involves four key design elements. The first is eliminating waste in our production and consumption processes so that waste from one production process (sometimes known as by-product) can be used in another productive process. Second, a restorative economy would encourage designing more labor into the process and using less energy and material resources (generally by reusing, reducing use, or recycling). Third, the restorative economy accounts for the true and full costs to our environment and communities in calculating profit. This generally requires a consideration of the lifecycle of a given product. Fourth, the restorative economy rewards producers with the least impact on the environment because their goods and services will be less expensive than those goods and services with a high impact on the environment.

Reference

Benyus, Janine M. 2002. *Biomimicry: Innovation Inspired by Nature.* New York: Perennial.

Carrying Capacity

The idea that the natural processes that support life have limits within which they can operate without failure suggests that excesses in population and/or consumption of resources can deplete our systems. The work of the Millennium Assessment supports this idea. Carrying capacity often tries to allocate population to some resources in order to make our relationship to finite resources more concrete.

Narratives about the relationship between human population and the limits of necessary physical support structures have captured the imagination. For example, Jonathan Swift proposed resolving Ireland's famine by eating dead children in "A Modest Proposal: For Preventing the Children of Poor People in Ireland from Being a Burden to Their

Parents or Country, and for Making Them Beneficial to the Public." Robert Malthus coined the word Malthusian to describe overpopulation in his famous essay, "An Essay on the Principle of Population." Garret Hardin describes how undervalued environmental resources are used to collapse in his article for *Science Magazine*, "Tragedy of the Commons."

There are dozens of ways to calculate a particular carrying capacity. Challenges to carrying capacity include how to evaluate the values of urban concentrations of infrastructure, and new ideas about spatial configurations and technology.

I = PAT

I = PAT is a formula that helps conceptualize the carrying capacity of an ecosystem. Impact on environmental systems is equal to the product of population, affluence, and technology. The more affluent a population, the more resources it tends to consume. Technology affects use of resources as well. Some technologies significantly increase the use of resources. For example, consumption can also be influenced by technology changes like the introduction and proliferation of personal computers and cars. Technological impacts may also be affected by the cultural and social behaviors such as privacy, conspicuous consumption, and individual ownership of personal property rather than group sharing. ***See also* Volume 2, Chapter 4: Tragedy of the Commons.**

References

Hardin, G. "Tragedy of the Commons." *Science* 162 (1968):1243–48. Online at www.sciencemag.org/cgi/content/full/162/3859/1243.

Manning, Robert E. 2007. *Parks and Carrying Capacity: Commons Without a Tragedy*. Washington, DC: Island Press.

Mitchell, Ronald B., William C. Clark, David W. Cash, and Nancy M. Dickson, eds. 2006. *Global Environmental Assessments: Information and Influence*. Cambridge, MA: MIT Press.

Changing How We Think about Economics

The way we think about a matter like the environment may be the most important factor determining the economic decisions we make. When we think about our surroundings as commodities and capital resources whose highest value is human use and financial profit, economic decisions tend to reflect short-term calculations and ignore longer term, systemic consequences.

Some have suggested that the way we think about a matter, the narrative that we tell ourselves about the issue and the mental paradigm we hold, is a far more effective method of change than any other method. For example, Donella Meadows, an expert in organizational development, says that the story we tell ourselves and our ability to change the story is the most effective method of changing our decision making and behaviors. In her theory, the least effective method of change is the approach of setting numerical standards and goals that many laws and management approaches adopt.

In economic terms, the narrative we have told about the relationship between nature and business is that nature is an endless, free set of resources intended to support and enhance human pleasures and welfare. The idea of commodities and natural resources is predicated on the separation of elements of nature from the systems in which they occur, and manipulation of them for human purposes. This separation and use ignore the systemic consequences of the uses to which we have put nature, and even worse in economic terms, has externalized onto nature the waste and pollution human use has caused. The challenging narrative of sustainability is one of biomimicry. If humans can model their interfaces and uses of nature with natural systems, it may be possible to live within the bounds of these systems and restore them rather than degrade them. Part of the shift in language traced in the following definitions parallels this attempt to recast the language and story of the relationship of economics to nature.

References

DiMento, Joseph F. C., and Pamela Doughman, eds. 2007. *Climate Change: What It Means for Us, Our Children, and Our Grandchildren.* Cambridge, MA: MIT Press.

Field, John. 2006. *Social Capital.* Andover, UK: Routledge.

Meadows, Donella. Leverage Points Places to Intervene in a System, www.sustainer.org/pubs/Leverage_Points.pdf.

Putman, Robert D. 2004. *Democracies in Flux: The Evolution of Social Capital in Contemporary Society.* London, UK: Oxford University Press.

Dematerialization

The concept of dematerialization involves using fewer materials and less energy to achieve comparable results in terms of functionality and consumer appeal. The challenge is to reduce the net environmental footprint of goods over their complete lifecycle. In terms of sustainability, this challenge would also include a complete accounting of the costs of production, not simply to the totality of materials used, but also the energy used to produce, use, and ultimately decompose or store goods and its effects on community. This idea is not unique to the green building movement. It has motivated innovative trends in architecture and design throughout the centuries, stimulating both enduring developments like the Roman arch and innovations like the geodesic dome.

Green Building

Businesses have led the way in architecture and industrial design that minimizes their environmental and energy footprint. By requiring such features in plants and other construction projects, businesses have created demand for what is called green building. This demand has been

Buckminster Fuller

"Bucky" is considered one of America's greatest visionaries of the 20th century. He used knowledge from architecture, visual arts, mathematics, molecular biology, environmental science, and literature to develop theories about how society could diminish the space between science and humanity. He is best known for his invention of the geodesic dome, the lightest, strongest, and most cost-effective structure ever constructed. The geodesic dome is able to cover more space without internal supports than any other enclosure, and the military praised it as the first major advancement in shelter in 1,600 years.

Fuller also was one of the first scientists to use renewable energy sources, such as wind and solar, in his inventions and ideas. He proclaimed, "There is no energy crisis, only a crisis of ignorance." He believed that our need for energy could be 100 percent satisfied by using wind generators attached to every high voltage transmission tower in the United States.

Bucky was also responsible for the diagram that shows the globe laid flat without surface distortion. It shows the world as one island and one ocean.

Reference

Buckminster Fuller: Starting with the Universe and Who is Buckminster Fuller—An Introduction to Buckminster Fuller at www.bfi.org.

R. Buckminster Fuller, www.worldtrans.org/whole/bucky.html.

FIGURE 2.7 • Desert Dome at Henry Doorly Zoo in Omaha, Nebraska. AP Photo/Nati Harnik.

magnified by governmental specifications in construction that also require reducing environmental and energy uses within buildings.

Businesses leave an enduring environmental footprint on their communities and environments in which they operate. Business environmental footprint includes consumption of natural resources such as fresh water, and it generates waste and pollution. So far, the more successful businesses have been, the greater this footprint, and the growth of trade through globalization has promised to enlarge this footprint. In the United States, an estimated 85 percent of our economy runs on nonrenewable fossil fuel, creating greenhouse gases, human health consequences, and waste. The majority of that energy usage is from a combination of industrial use, commercial and institutional buildings, and commercial transportation. A substantial minority of our energy usage occurs in our households and individual transportation needs such as commuting.

Buildings use the largest portion of our material resources, including energy. Buildings are the single largest contributors to factors that create waste and pollution because of the uses of energy that occur within them and the materials they are composed of. Careless or inefficient uses of energy to heat, cool, or light buildings is another area of considerable energy waste. Intentional and transient or phantom uses of energy within buildings consume considerable amounts of energy. Building material fills landfill space far more rapidly than household waste. The green building movement engineers waste out of the lifecycle of buildings by eliminating inefficient uses of energy, reducing and reusing materials.

Construction of the built environment is undergoing a radical change in approach from concept to cement, and everything in between is being reconceptualized. Technology, both old and new, is being redeployed to save energy. Spatial relationships and needs are being reconfigured to eliminate the elements that lead to sprawl, dependence on fossil-fueled transportation, and social alienation and isolation.

Businesses have led the way in architecture and industrial design that minimizes their environmental and energy footprint. By requiring such features in plants and other construction projects, businesses have created demand for what is called green building. That demand has been magnified by governmental specifications in construction that also require reducing environmental and energy uses within buildings.

LEED Certification

LEED certification is one green building program designed and supervised by the U.S. Green Building Council. LEED sets energy and environmental standards for the design and construction of all sorts of buildings, including new and existing structures. LEED stands for Leadership in Energy and Environmental Design. The U.S. Green Building Council is a nonprofit organization that promotes sustainability in the built environment. In 2000, this group endorsed a series of construction standards applicable to buildings whether they are new construction or remodeled, residential or commercial. These standards are aimed at achieving lower impacts in terms of energy use and environmental footprint. Buildings that satisfy these standards can be awarded a LEED certificate based on their accomplishments. LEED training and certification are now administered by the Green Building Certification Institute.

This certification responds to a growing concern that claims of sustainable design and products are being misused in commerce, a practice some call "greenwashing."

References

Cole, Raymond, and Richard Lorch. 2003. Buildings, *Culture and the Environment: Informing Local and Global Practices*. Hoboken, NJ: Wiley-Blackwell.

Yudelson, Jerry. 2008. *Green Building through Integrated Design*. New York: McGraw Hill Professional.

When structures achieve energy efficiency, and when building materials do not become waste, giant strides are taken to reducing human ecological footprints. Green building is a trend in the design of built materials and structures that explores ways to minimize the impact of buildings on their ecology. Green building techniques try to achieve energy efficiency and to use recycled materials and/or ensure that building materials can be recycled in the future. It also tries to eliminate runoff water pollution and use renewable sources of energy.

LEED (Leadership in Energy and Environmental Design) certification is one green building program designed and supervised by the U.S. Green Building Council. LEED sets energy and environmental standards for the design and construction of all sorts of buildings, including new and existing structures.

Other ways in which businesses have begun a new relationship with their environmental impact include industrial ecology and product lifecycle management. Innovations in these areas attempt to design waste and pollution out of production processes, while still using an industrial model of production. ***See also* Volume 2, Chapter 3: State and Local Initiatives.**

Gross Domestic Product: Gross Domestic Happiness

One of the ways we measure the performance of an economy is to add up the total value of all the goods and services that it produces in a year. This is called the gross domestic product. One of the criticisms of this measurement of an economy's performance is its failure to distinguish between goods and products that denote a happy or constructive condition, and those that thrive on misery. For example, if a society is experiencing unhealthy conditions, the value of its expenditures on medical care may increase dramatically, especially emergency care. Although this condition would add to the gross domestic product, it does not indicate whether a population is healthy or happy.

King Jigme Singye Wangchuck of Bhutan, and Gross National Happiness

The fourth dragon king of Bhutan ruled his country from 1972–2006. He voluntarily relinquished power to a form of democracy in 2008. He is credited with the term gross national happiness (GNH), by which he chose to measure his country's developmental progress.

There are continuing conferences and theories of how to measure gross national happiness. The value of economic activities are included based on their contribution to well-being, not simply dollar values. The difference this makes can be seen in comparing the dollar value of an involuntary hospital stay (presumably unhappy and not included in GNH) with the dollar value of starting a new business.

Industrial Ecology

Industrial ecology is a discipline that combines features of engineering and other sciences to eliminate toxic materials and waste from industrial processes and products. Key concepts that are associated with this discipline include extended manufacturer responsibility for products throughout the lifecycle of a product, using fewer energy and material resources to produce more, and pairing or grouping businesses so that waste from one becomes a material resource to another.

Contemporary engineering often sees the production of products as a straight line from resources to finished product. The addition of other considerations, including the impacts of production on environmental conditions and social conditions, broadens the concept of design to include ecological factors. Industrial ecology is a science devoted to designing industrial processes in ways that coexist with or restore natural systems and communities. It uses a multidisciplinary approach to industrial processes intentionally to eliminate waste, pollution, and inequities. ***See also* Volume 2, Chapter 2: Language of Sustainable Business.**

Intelligent Product Systems

Designing waste and pollution out of new products and their packaging can be done by producers who accept responsibility for the full lifecycle of a product. Intelligent product systems thinking make use of those distinctions by imposing different responsibilities on sellers of goods that will not decompose readily, and goods that release toxic and hazardous materials. Intelligent product systems often make manufacturers of durable goods responsible for the ultimate depreciated product. This provides manufacturers with an incentive to reuse these materials or search for ones that decompose into soil reasonably quickly. Intelligent product systems would impose a fee for the life of these toxic and hazardous materials on the manufacturer who uses them in the manufacturing process or product. That fee would have two effects: (1) to discourage the use of these materials and encourage the search for

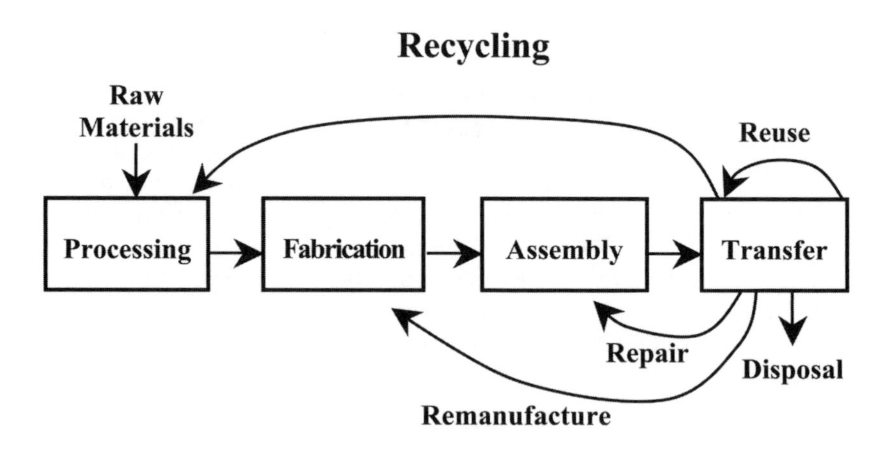

FIGURE 2.8 • Industrial ecology is a design system built on mimicking the way natural systems work.

nontoxic alternatives and (2) to help the government pay for the costs and consequences of toxins and hazardous waste.

Current legislation in Germany and the European Union is close to adopting these principles for materials used for packaging.

References

Hitchcock, Darcy, and Marsha Willard. 2006. *The Business Guide to Sustainability: Practical Strategies and Tools for Organizations.* London, UK: Earthscan.

Hunkeler, David et al. 2008. *Environmental Life Cycle Costing.* Boca Raton, FL: CRC Press.

Seliger, Gunther. 2007. *Sustainability in Manufacturing: Recovery of Resources in Product and Material Cycles.* New York: Springer.

Stern, Alissa J. 2000. *The Process of Business/Environmental Collaborations: Partnering for Sustainability.* Westport, CT: Greenwood Publishing.

United Nations Environmental Program. 2007. *Life Cycle Management: A Business Guide to Sustainability.* New York: UN Publications.

Yanful, Earnest K. 2000. *Appropriate Technologies for Environmental Protection in the Developing World.* Boca Raton, FL: Springer Press.

Natural Capitalism

Natural capitalism refers to the idea that capitalism can function as a restorative economy if the benefits that nature provides are valued and accounted for as capital assets. Market mechanisms used by capitalism to develop unsustainably are thought to be capable of generating sustainable behaviors when true and full costs of production are factored into market transactions.

This idea posits that as resources become more expensive through true and full cost accounting, resources will be more valuable than the cost of labor. Under these conditions, there will be an incentive to minimize resource use and maximize the use of labor. In their work, Paul Hawken and Hunter and Amory Lovins call for a resource revolution in terms of efficient use of resources and maximum use of labor.

Reference

Hawken, Paul, Amory Lovins, and L. Hunter Lovins. 1999. *Natural Capitalism: Creating the Next Industrial Revolution.* Boston: Little, Brown. Book chapters online at www.natcap.org/sitepages/pid5.php.

Product Lifecycle

Products all have a beginning, middle, and end. Their beginning reflects their new state, undiminished by use. Producers are responsible for new products in terms of the costs of production and profits. As a product is used, it is consumed, although different kinds of products are consumed at different rates. This is also called the process of depreciation. When a product is sold, the consumer becomes responsible for its use. When a product's useful life is over, or when it is fully depreciated, its lifecycle ends.

Products move through a cycle from conception or design through manufacture, sale/use, and ultimately decomposition. Human involvement at each stage has usually not been responsible for more than a small part of that product lifecycle. Lack of comprehensive responsibility has led to lack of awareness from designers through ultimate consumers about the consequences of waste and pollution from these products. For example, computers have transformed our contemporary society in terms of some fundamental measures of productivity. Their designs have become more compact as miniaturization of their components has increased, but their production consumes more and more material resources using more energy, and their disposal creates vast amounts of waste.

When the useful life of a product is over, the challenge of sustainability begins again. A used product often becomes waste, and it can add to pollution present in watersheds, air sheds, and landfills. A product may be designed to be reusable in a new product or biodegradable, eliminating waste. Producers who design their products in this way eliminate waste. Some describe this kind of design as biomimicry or a closed loop. Sometimes producers are required to incorporate this type of design by law.

Important differences exist in terms of the environmental consequences of their lifecycles between items that will decompose into soil relatively quickly, those that will not decompose quickly at all, and those whose decomposition will release toxic and hazardous materials into the environment.

Products that decompose quickly become the basis for further productive cycles. This is one important means of eliminating waste and also imitating nature's own closed loops. Goods that do not degrade into soil quickly are often called durable goods because of their longevity. Types of durable goods include cars, refrigerators, washing machines, and computers. When these types of products enter a landfill, they will remain there indefinitely, adding to the burden on land and water resources and releasing further toxic materials. The materials in these products are often valuable but present in such complicated arrays that harvesting these items from landfills is technically difficult. For example, Japan's landfills now hold enough gold, silver, platinum, and indium to make it as resource-rich as countries in the top five sources of these resources because Japan discards electronic consumer gadgets like cell phones, media players, and computers at an extremely high rate. If manufacturers become responsible for the entire lifecycle of these products, they would have more of an incentive to harvest and reuse such valuable materials before they end up in a landfill.

Finally, hazardous and toxic materials are dangerous to the environment and the communities where they come to rest. Against statistical probability, these materials come to rest most often in communities of people of color and poor communities. This observable phenomenon mobilized the environmental justice movement in the United States and worldwide. The consequences of these materials to human health and the environment is enormous if they are not contained and treated. The costs to contain and treat them is expensive. Yet, businesses that use these materials to make profits do not pay these costs and rarely pay for these consequences. Economists call this an externality, when these costs are borne by others, not the maker.

Businesses are interested in designing waste and pollution out of their production processes and products because this improves profits, and they are often motivated by the desire to do what is good for the environment. There are many approaches to achieving these results. The U.S. government has sponsored educational projects on this topic for businesses such as Design for the Environment. Universities have developed interdisciplinary specialties in industrial ecology.

Renewable Resource/Nonrenewable Resources

Our economies use resources in the sustenance of human communities and enterprises. All of these resources come from natural processes and natural elements. Some of these resources cannot be replicated and are not re-created on a predictable basis. For example, fossil fuels are materials used to power our basic needs for energy. These fuels come from the fossilized remains of animals and plants that were alive during prehistoric times and are now extinct. The supply of these remains is limited. Once used, their remains will never be replaced. These are an example of nonrenewable resources. To the extent our economies and businesses rely on these types of resources, they are not sustainable.

Renewable resources are those resources that are re-created and replaced by natural processes at predictable intervals. For example, sunlight is a powerful force governing and regulating life on Earth. We predict its presence and absence every day in the calculation of daylight and nighttime, as well as weather forecasting.

The rate at which human enterprises use renewable resources matters because use that exceeds the capacity of natural processes to replace them is unsustainable use of renewable resources. The rate at which we use nonrenewable resources matters even more because when such resources are used up, they are gone forever. Sustainability requires us to maintain our enterprises in ways that do not diminish the ability of future generations to maintain their own communities and enterprises.

When resources are used, they do not disappear. Use of resources generates other resources as those used dematerialize into constituent parts. In nature, all these constituents become the basis of new processes. In natural cycles of use, there is no waste. That is sometimes conceptualized as a closed loop. The process of dematerialization or transmaterialization during which materials become something else through thermal and chemical processes is fundamentally different from the idea of garbage or trash in which material is discarded without future foreseeable use. ***See also* Volume 2, Chapter 4: Energy; Volume 2, Chapter 4: Energy Sources.**

Restorative Economy

The idea that human enterprise could be organized along the lines of nature without open loops is the beginning of the idea of organizing human economic enterprise in a sustainable way. Some visionary economists and business people argue that human economies could be organized in a way that facilitates the renewal and restoration of natural processes. Increasing profits by decreasing environmental and social impacts is the core of a restorative economy. Profit would be allocated to the producer whose goods and services have the least impact on natural processes. In that way, competing for goods and services that have the least impact on natural processes of restoration is one way that economic competition could restore natural processes.

For profit to be related to impact on natural processes, resources would have to reflect their full and true costs, and producers would have to pay the true and full costs of the resources used. In economies distorted by subsidies and externalities, resources are not reflective of their value in natural processes and competing for profits based on the lowest price escalates the generation of waste and pollution.

One important cornerstone of a restorative or sustainable economy is valuing natural systems and acting as if those systems have economic values. The United Nations Green Economy Initiative has created a project to establish the economic value of ecosystems services like those identified in the Millennium Ecosystem Assessment. TEEB—The Economics of Ecosystems and Biodiversity—is the name of a two-year, two-phase project sponsored by the G-8+5 nations and the United Nations Environment Program. The goal of the study is to motivate actions to reduce significantly the loss of ecosystem services by 2010. The interim report of the TEEB states that a "business-as-usual" approach to the degradation of ecosystems will have grave consequences by 2050.

References

Hawken, Paul, Amory Lovins, and L. Hunter Lovins. 1999. *Natural Capitalism: Creating the Next Industrial Revolution*. Boston: Little, Brown. Book chapters online at www.natcap.org/sitepages/pid5.php.

TEEB—The Economics of Ecosystems and Biodiversity, www.unep.org/greeneconomy/index2.asp?id=teeb.

Triple Bottom Line

Some corporations have tried to integrate sustainability into their business practices by accounting for more than simple profits. In addition to their financial accounting, these businesses also account for the social and environmental impacts of their enterprise.

Some businesses question the need for their additional accountability for fairness or equity to the workers or the communities in which they operate, even though they may embrace responsibility for the environmental consequences of their business conduct. The triple bottom line (TBL) means that the business organization embraces financial, environmental, and social values; profits are no longer exclusive. In addition to economic value, the TBL approach focuses corporations on the environmental and social values they add, or destroy, in society. The essence of TBL accounting involves the expansion of the corporate reporting framework, traditionally focused solely on financial capital, to include natural and human capital in the bottom line equations. Traditional accounting is shortsighted, measuring in 12-month accounting cycles or quarterly measurements. TBL practice requires a long view, encompassing generations into the future and taking into account externalities—social and environmental costs not recorded in accounts—such as the impact building a new plant in a particular neighborhood will have on property taxes.

- Economic Bottom Line: Accountability for Profits

A company's bottom line is the net profit or earnings—the gross profit minus all expenses. Traditional capital comes in two forms: physical (e.g., machinery and plants) and financial. The move toward TBL paradigm requires the inclusion of human capital (experience, skills and knowledge-based assets) and intellectual capital (employee competencies, work processes, trademarks, customer lists, for example).

- Environmental Bottom Line: Accounting for Environmental Consequences

The environmental bottom line involves the responsible use of natural capital and attention to minimizing the impact of waste and pollution on the planet. Measurement of natural capital goes beyond counting board feet of timber in a forest to account for the underlying wealth stream that supports the forest ecosystem. That stream of wealth produces timber and other commercial products, but also absorbs greenhouse gases and filters ground water, supporting local flora and fauna, and commercial fisheries.

One critical data point for assessment is resource use: efficient and managed consumption of resources, including water, electricity, purchased end-user products like office supplies; reducing waste; asserting responsibility for safe disposal of hazardous materials. Natural capital's bottom line requires a life cycle assessment of products to capture the real costs: from the raw material to manufacture to distribution to disposal. Such externalities are left out of traditional accounting but are necessary to an understanding of the whole.

- Social Bottom Line: Accounting for Communities

Social capital encompasses human capital in the form of public health, skills and education. Basic assessments include treatment of labor such as providing fair salaries and benefits, and safe work environment, and the level of long-term commitment to employee advancement and success. Beyond that is a concern for the larger community in which the corporation operates. The social capital of a new plant includes the plant's potential impact on the health of existing and future residents surrounding the proposed site. A company that acknowledges the interconnectedness of corporation, employees and community stakeholders is in a stronger position to grow sustainably.

Social capital also includes a wider measure of capacity—human potential for social health- and wealth-creation. The ability of people to work together for common goals is critical to this level of social capital, and it is essential to building a sustainable society. The degree of trust between business and external stakeholders will be a key factor to their long-term sustainability.

References

Field, John. 2006. *Social Capital.* Andover, UK: Routledge.

Putman, Robert D. 2004. *Democracies in Flux: The Evolution of Social Capital in Contemporary Society.* London, UK: Oxford University Press.

True and Full Cost Accounting and Pricing

The webs of life in nature provide humans with many invaluable, irreplaceable benefits.

> Try to imagine the technologies that could replace the services nature provides, including these noted by Dutch scientist Rudolf S. de Groot: Oxygen production of water and air; regulation of atmospheric chemistry; protection against cosmic and ultraviolet radiation; solar energy; regulation of local and global climate; maintenance of biological and genetic diversity; maintenance of wildlife migration and habitats; storage, detoxification, and recycling of human waste; natural pest and disease control; genetic and medicinal resource production; prevention of soil erosion; sediment control; regulation of runoff and flood prevention; regulation of the chemical composition of oceans; formation of topsoil and maintenance of soil fertility; production of fertilizers and food; nutrient storage and recycling; and fuel and energy production.
>
> Paul Hawken, 1997, p. 42

These benefits are impaired and dissipated by overuse, waste, and pollution. When these benefits are not reflected in the price of goods and services sold, the market undervalues them. Goods and services will then be cheaper than they should be if these benefits were recognized and included in costs. Undervalued resources are subject to waste and ultimate depletion. Full value of these benefits would encourage sustainable behavior by discouraging waste and pollution.

References

Costanza, Robert, and Lisa Wainger. 1991. *Ecological Economics: The Science and Management of Sustainability.* New York: Columbia University Press.

Daly, Herman E., and Joshua Farley, eds. 2009. *Ecological Economics: Principles and Applications,* 2nd ed. Washington, DC: Island Press.
Hawken, Paul. "Natural Capitalism." *Mother Jones Magazine* March/April (1997):42.
Repetto, Robert C. 1986. *World Enough and Time: Successful Strategies for Resource Management.* New Haven, CT: Yale University Press.
Schaltegger, Stefan et al. 2006. *Sustainability Accounting and Reporting.* New York: Springer.

ATMOSPHERE

Earth's atmosphere is approximately seven miles of air surrounding the planet and extending into outer space. In the atmosphere water and air interact with each other and with the land below.

Air can hold water in its gaseous form until temperature causes water to condense. Condensing water can fall from the sky to Earth in many forms of precipitation.

The atmosphere is a very dynamic air mass, integrally related to land and water bodies. Environmental degradation of the air can result in environmental degradation of land and marine based ecosystems. Environmental degradation of the atmosphere is also becoming increasingly observable with the increase of satellite monitoring. Natural events such as volcanic eruptions can load the atmosphere with so much particulate matter they darken large parts of Earth. Human impacts on the atmosphere, such as air pollution, ozone depletion, and overdevelopment, can also load the air with particulate matter. This can cause air inversions that trap poisonous air in communities. Because the atmosphere is so integrally related to land and water systems, this particulate (and other chemicals) matter eventually affects land and water as well.

The ecosystem impacts of atmospheric degradation greatly concern sustainability advocates. Environmental monitoring of the atmosphere is now able to locate many more major sources of air pollution.

References

Gunn, Angus M. 2003. *Unnatural Disasters: Case Studies of Human Induced Environmental Catastrophes.* Westport, CT: Greenwood Press.
Volk, Tyler. 2008. *CO2 Rising: The World's Greatest Environmental Challenge.* Cambridge, MA: MIT Press.

Fossil Fuels

Humans burn materials formed during prehistoric periods as sources of energy for most of their basic needs. These fossilized materials are tens of thousands of years old and must be taken from the crust of the Earth. They are rich in hydrocarbons, a type of complex molecule formed of hydrogen and carbon. When these materials are burned, they give off energy in the form of heat, gases, and particle matter that enter the atmosphere. These gases include greenhouse gases that can cause Earth's atmosphere to become warmer. These particles include volatile organic compounds that can cause danger to human health.

Petrochemicals are now used in the production of many goods. Many plastics, cosmetics, and building materials have petrochemical components. Petrochemical dependence is a motivating force for sustainability because many nations strive to be energy independent. Reliance on another country for a nonrenewable resource, like gas and oil, often stifles the economic development and political independency of that nation. The search for alternative energy sources could result in more sustainable approaches.

Volatile Organic Compounds

Some chemical compounds are able to be vaporized and can enter the atmosphere in the form of gasses. These can happen at room temperature or in the range of temperatures of many industrial processes. As these chemicals volatize, they can present a threat to workers, community residents, and the environment depending on the toxicity of the chemical and the exposure to it.

Volatile organic compounds represent one of the early, big steps in environmental policy development around air quality. They brought in a range of volatizing chemicals, and there is an environmental and industry debate on almost every one. Environmentalists and most sustainability advocates claim more chemicals should be regulated.

Greenhouse Gases

When fossil fuels are burned, they produce energy and gases. Some of the gases released into the atmosphere have the effect of causing chemical and thermal reactions in our atmosphere. Nitrous oxide is a greenhouse gas that has been linked to acidification rain and ocean. Carbon dioxide is another greenhouse gas linked to climate change.

Most of the increase in global average temperatures since the mid-20th century is likely due to increases in the emissions of gases caused by human activities. Global greenhouse gas emissions have grown 70 percent from 1970 to 2004. Carbon dioxide emissions have grown about 80 period in the same period.

One of the biggest concerns many sustainability advocates have regarding the atmosphere is the effect of carbon dioxide on the erosion of the ozone layer. The ozone layer protects the Earth from harmful ultraviolet radiation from the sun. A major part of the ecological footprint of cities and organizations is their carbon emissions. Many organizations and meetings now arrange it so it is possible to buy carbon offsets. Carbon offsets can be somewhat controversial because their efficacy with regard to carbon emissions is questionable. Others have criticized them because their volume is just too small to make a difference if done voluntarily. There are other proposals of carbon dioxide sequestration

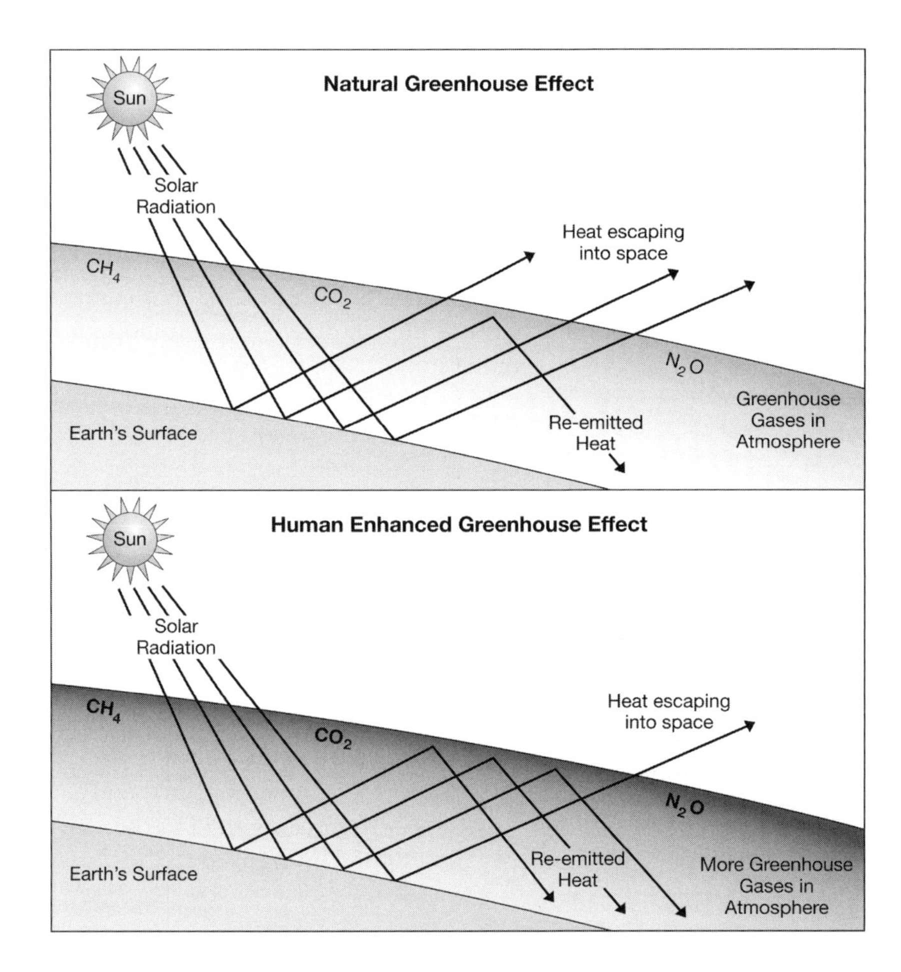

FIGURE 2.9 • How greenhouse gases trap heat in the Earth's atmosphere. Illustrator: Jeff Dixon.

that would theoretically pump carbon dioxide emissions into underground storage areas. These areas would have to be very large, and the amount of energy to move that amount of carbon dioxide might itself have negative environmental impacts. Some more recent research indicates that it may be possible to sequester large amounts of carbon dioxide in a type of rock called peridotite. This type of rock removes carbon dioxide from the air and stores it as an inert chemical as limestone and other carbonates. The process would work by pumping carbon dioxide down bore holes hundreds of feet deep into peridotite. The peridotite would be exploded to create more surface area for the carbon dioxide, and heat would be added to begin the chemical action of making the carbon dioxide inert. Scientists speculate that once started, the fire would continue on its own. Some areas of the world have large areas of peridotite. The nation of Oman has approximately 2,400 square miles that could sequester about 4 billion tons of carbon dioxide a year.

Reference

Volk, Tyler. 2008. *CO2 Rising: The World's Greatest Environmental Challenge.* Cambridge, MA: MIT Press.

Climate Change or "Global Warming"

As the balance of gases in the atmosphere tips toward greenhouse gases, the Earth's atmosphere holds more heat, and as the temperature of our atmosphere rises, climate zones around the world are changing. The causes of climate change includes human introduction of greenhouse gases into the atmosphere. How much of climate change is human driven and the speed of climate change are controversial. Scientists agree that climate change is occurring, and much of it is human driven or anthropogenic.

Climate changes are of extreme importance to sustainability advocates because it provides evidence of the limitations of Earth's natural systems. The evidence of climate change affects water, ecosystems, food, coasts, and health of land animals, especially humans. The Intergovernmental Panel on Climate Change summarized its predictions of the impacts on these systems by degree of global mean annual temperature change in the accompanying figure. ***See also*** **Volume 2, Chapter 4: Climate Change.**

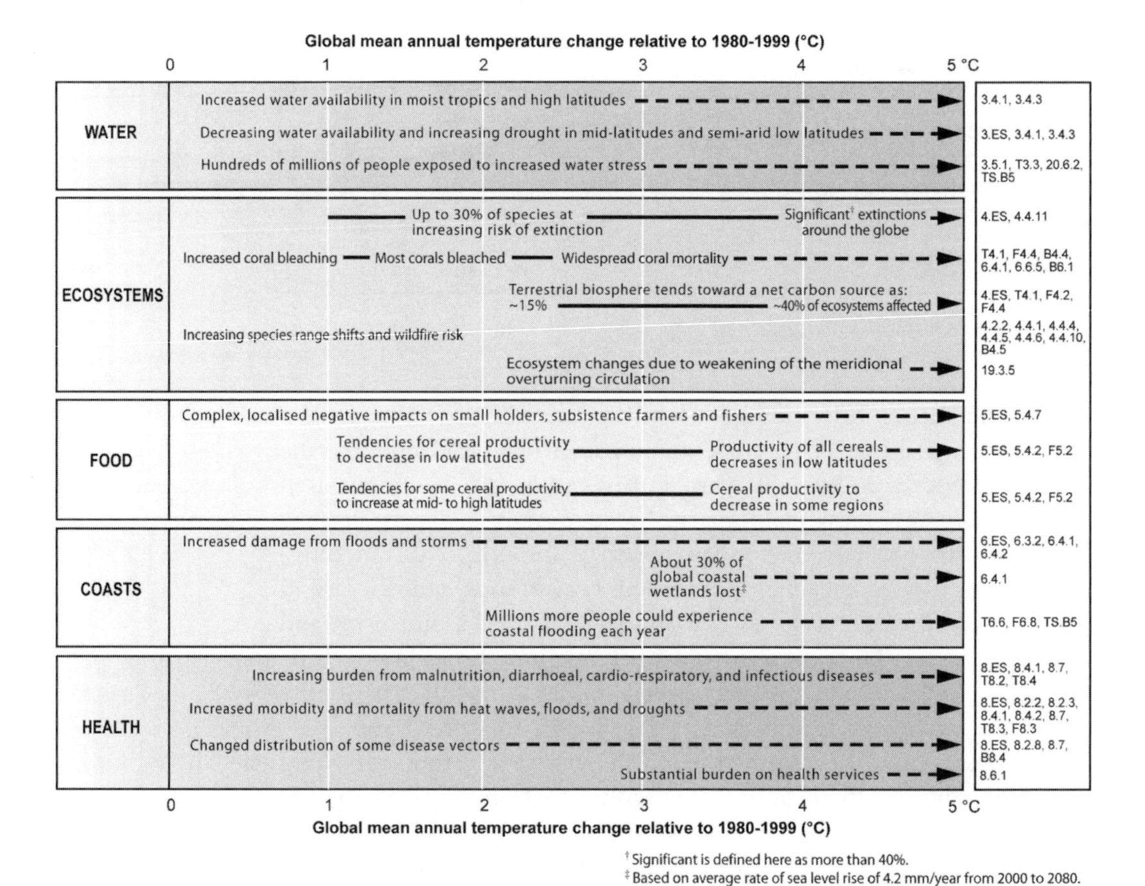

FIGURE 2.10 • Key Impacts on Natural Systems as a function of increasing global average temperature change. Reprinted from Intergovernmental Panel on Climate Change, Climate Change 2007—Impacts, Adaptation and Vulnerability: Working Group II contribution to the Fourth Assessment Report of the IPCC (Geneva, Switzerland: Intergovernmental Panel on Climate Change, 2008), 10.

Water impacts of global warming is predicted to increase water availability in moist tropics and higher latitudes, decrease water availability and increase drought in mid- latitudes and semiarid low attitudes, and expose hundreds of millions of people to water stress. Ecosystems are predicted to increase species extinction with increase in temperature, coral bleaching and mortality, and wildfire risks. In terms of food sheds, global warming is predicted to cause tendencies for cereal productivity to decrease in low latitudes and for some increase in cereal productivity to increase in mid to high latitudes, unless it gets warmer faster; in that case, productivity will decrease in all latitudes. Rising ocean levels from global warming will be a significant part of climate change. It is predicted that damage from floods and storms will increase, with the disappearance of as much as 30 percent of the wetlands. In terms of health, there will be an increasing burden from malnutrition and diarrheal, cardiorespiratory, and infectious diseases; increased morbidity and mortality from heat waves, floods, and droughts; and changes in the distribution of some disease vectors.

As more evidence mounts of the limitations of Earth's natural systems sustainability, proponents hope this translates into meaningful approaches to public policy and private behavior changes toward sustainability.

References

Dimento, Joseph F. C., and Pamela Doughman. 2007. *Climate Change: What It Means for Us, Our Children, and Our Grandchildren.* Cambridge, MA: MIT Press.

Emanual Kerry. 2007. *What We Know about Climate Change.* Cambridge, MA: MIT Press.

Parry, M. L., O. F. Canziani, et al., eds. 2007. *Climate Change 2007: Impacts, Adaptation and Vulnerability. Contribution of Working Group II to the Fourth Assessment Report of the Intergovernmental Panel on Climate Change.* Cambridge, UK: Cambridge University Press.

BIODIVERSITY

The three types of biological diversity—plants, microorganisms, and animal life—are the subject of ongoing study. We do not know how many types of life Earth hosts, even while we are losing some species every day as a result of planetary changes. This is one type of biological diversity. Another type of biological diversity is the range of ecosystems. A third type of biological diversity relates to the genetic diversity within a particular species. Biodiversity occurs with both wild and domesticated animals. Most cattle breeds that are currently used for dairy and meat production are from a wild ox called the Aurochs. It was six feet tall and the last one died in Poland in 1627. In all, 1,491 of the 7,616 domesticated breeds of animals are in danger of becoming extinct; 9 percent are now extinct.

The loss of biological diversity relates to losses in all three types. They are a great concern for advocates of sustainability with an ecological focus because they are occurring at many times the natural rate. According to one United Nations report, more than 3,000 wild species showed a consistent decline of 40 percent from 1970 to 2000. Inland, generally freshwater species declined in population about 50 percent. Other species declined about 30 percent. The same report concluded that between 12 and 52 percent of bird species could be extinct. The main reason is habitat loss, especially the loss of tropical rainforests. Invasive species also cause loss of biodiversity by displacing native plants. Climate change may also increase loss of biological diversity through species loss. Some species may not be able to adapt to rapid changes in climate. The environmental impacts on the largest habitat on Earth, the deep ocean, are still unknown.

This loss of biological diversity has large implications for ecologically based sustainability advocates. In some ways, it politically motivates individuals and governments to take stock of their environment so they can consider how to act sustainably. Some scientists project that the loss of genetic and species biodiversity is already well on its way. The main characteristic of surviving species is how well they get along with humans. An unknown and controversial aspect to preserving all types of biological diversity is the use of technology. Another site-specific question is how much biodiversity is necessary to support the local ecosystem or biome? How much carrying capacity does the local ecosystem have?

A fundamental policy question for ecologically based sustainability proponents is how much biological diversity is necessary and what kind of biological diversity do we want to sustain.

References

Bachmann, Peter, Michael Kohl, and Risto Paivinen. 1998. *Assessment of Biodiversity for Improved Forest Planning.* New York: Springer.

McNeill, J. R. 2000. *Something New Under the Sun: An Environmental History of the Twentieth Century World.* New York: W. W. Norton.

O'Riordan, Timothy, and Susanne Stoll-Kleemann. 2002. *Biodiversity, Sustainability and Human Communities: Protecting Beyond the Protected.* Cambridge, UK: Cambridge University Press.

Stein, Bruce A. et al., eds. *Our Precious Heritage: The Status of Biodiversity in the United States.* New York: Oxford University Press, 2000.

United Nations Secretariat of the Convention on Biodiversity. 2006.*Global Biodiversity Outlook.* Montreal: Secretariat of the Convention on Biodiversity.

FISHERIES AND FISHING

An early warning of how large human impacts can be relates to the depletion of marine fisheries. Humans have always fished. Fishing is a way of life, recreation, and spiritual need. Technology has so changed human proficiency at fishing that Jacques Cousteau, the legendary ocean explorer, described it as a transition from hunting to harvesting. Large ocean trawlers with nets drifting for miles and catching everything in its path can quickly deplete fishing reserves. Good fishers had to know where the fish were, but radar and other fish-finding devices assist in that with efficiency.

Fish population counts, enforcement of fishing regulations domestically and internationally, and strongly held fishing traditions all combined to create an area of controversy. The marine fishery populations have declined by such large numbers, however, that the evidence of severely negative environmental impacts from overfishing, underenforcement of domestic and international fishing regulations, and cultural rigidity is indisputable. Fishing populations have all generally declined. Large predator fish such as tuna have declined by 90 percent over original fishing stocks. Human impacts are decimating many other marine environments. Coral reefs and large groves of mangroves are negatively impacted by excessive fertilizer and silt runoff. About one-fifth of coral reefs globally are gone, and mangroves have suffered large and immeasurable loss of habitat.

Species of fish are generally found in the environment in specific localized areas at certain times, unless they are farmed or stocked by humans. The stock of such locally found species of fish are called fisheries. Humans hunt these stocks for food and sometimes for fuel and fiber. Fisheries are sometimes described as a type of commons in that everyone shares a common resource that is irreparably damaged if too much is taken. Some groups of fishermen have sought to be sustainable. In September 2007, the San Diego American Albacore Fishing Association became certified as sustainable by the Marine Stewardship Council. The Marine Stewardship Council is a nonprofit organization that certifies that food is caught in ways that avoid overfishing. ***See also* Volume 2, Chapter 4: Tragedy of the Commons; Volume 2, Chapter 5: Shares.**

References

Hannesson, Rognvaldur. 2006. *The Privatization of the Oceans.* Cambridge, MA: MIT Press.

Webster, D. G. 2008. *Adaptive Governance: The Dynamics of Atlantic Fisheries Management.* Cambridge, MA: MIT Press.

Habitat

The kind of plant life, water systems, and other ecological configurations necessary to support a particular plant or animal's life is called its habitat. It is a physical place that is necessary for a plant or animal to live and procreate. It can be more than one place, as in the case of migratory species like some fish and birds. It is the place the plant or animal lives.

The loss of many plant and animal species is often attributed to loss of habitat. When wilderness areas are opened up for oil drilling, as in the case of the Arctic Wildlife Refuge, the habitats of migrating species such as the caribou is damaged. Species that follow the caribou, such as wolves and mosquitoes, are also affected by habitat loss. A total of 15,000 to 30,000 species extinctions occur because of human activities and their contribution to habitat loss. Aquatic species are also affected by habitat loss. In the United States about 20 percent of freshwater fish species and about 50 percent of mollusks are either extinct or are threatened with extinction because of habitat loss.

Human population growth and its increased impact on the environment are a direct cause of habitat loss. Human actions like agribusiness, clear-cut logging, massive mining operations, water control systems like dams and diversions, building construction, and road construction all damage habitat. Animal and plant species that can no longer find food and water in these former habitats become threatened with extinction.

Habitat preservation is one of the main reasons for the development of state and national parks. Wilderness areas stress habitat preservation, and their advocates resist all mining, oil drilling, logging, and road development. The protection of endangered species in wilderness areas is controversial because wilderness areas are designed to protect the habitat of threatened species, even unpopular ones that can cause economic loss to nearby property owners. An example of this is the protection of wolves in national parks.

References

Meyer, Stephen M. 2006. *The End of the Wild.* Cambridge, MA: MIT Press.

O'Riordan, Timothy, and Susanne Stoll-Kleemann. 2002. *Biodiversity, Sustainability and Human Communities: Protecting Beyond the Protected.* Cambridge, UK: Cambridge University Press.

Animal Language, Communication, and Music

Most current scientific studies of biodiversity examine population counts. Sometimes they examine migratory paths, their niche in a given ecosystem, and their evolutionary pathways. They seldom examine the way animals learn, communicate, and think. How animals communicate may affect their ability to survive, which affects their population.

Loud noises increase stress levels for most animals, including humans. Noise affects how animals react to their environment. Human settlements, with airports, sirens from emergency vehicles, and high background noise, generally drive animals away. Loud noise in nature can mean danger, especially if it is a strange or unusual noise. It can affect how animals breed, migrate, and find food and shelter. If these activities are hindered, the animal populations tend to dwindle. When the numbers of animals decrease, other aspects of the ecology can be affected. Plants dependent on certain animals for procreation, such as pollen carried by bees and butterflies, may be negatively affected. The overall biodiversity of a given ecosystem suffers, and if it suffers irreparable damage, then chances for sustainability are nil.

Research into the language and music dimensions of animals is just beginning. It is also very controversial, with some linguists maintaining that language is the sole area of humans. The language abilities of animals depend on the animal. Whales and elephants are known to communicate long distances using infrasonic sound imperceptible to humans. Dolphins and birds communicate using a variety of sounds. Research on prairie dogs shows that they communicate using key aspects of language—nouns, verbs, and adjectives. Prairie dogs inhabit the same colonies for many years. They are the favorite prey of almost every hunter in their ecosystem. Prairie dogs communicate about individual coyotes who hunt them. They communicate with each other about whether the human hunters have a gun. They can communicate with each other about the hunting activity of specific hawks.

Animal Music

The differences and similarities between language and music are complex. Music tends to have a relationship between sounds that form certain ratios that form patterns. Music is made up of melody, rhythm, meter, tonality, and timbre as the primary patterns. Some languages such as Mandarin Chinese are tonal in character. Some researchers believe that animal music may predate human music because whales predate humans. Only about 10 percent of all primate species make music. Some whale species use songs that repeat refrains and that rhyme. Birdsongs have intrigued humans for centuries and have distinct meanings. The famous music composer Mozart kept a starling as a pet. He recorded a part of the Piano Concerto in G Major as his pet starling had revised. The bird "revised" it by changing the sharps to flats. Mozart's composition, "A Musical Joke" was written the way a starling sings music. Birds use music very much like humans beyond simple mimicry. Birds use sonatas, variations in rhythm and pitch, accelerandoes, crescendos, diminuendos, and human sound scales.

In terms of biodiversity, recognizing animal communication, whether it be language or music, is necessary to understand animals and their relationship to systems of nature on which future life depends.

References

Gray, P. M. et al. "The Music of Nature and the Nature of Music." *Science* 291 (2001):52–54.

Slobodchikoff, C. N. 2002. "Cognition and Communication in Prairie Dogs." In *The Cognitive Animal: Empirical and Theoretical Perspectives on Animal Cognition,* eds. Marc Bekoff, Colin Allen, and Gordon M. Burghard, 257–64. Cambridge, MA: MIT Press.

Monoculturing

Some varieties of plants and animals become popular at a particular time because of their appeal to businesses and consumers. Some varieties of plants store well for extended periods, making them better for long-range shipping. These preferences may take over the cultivation of these types of foods, resulting in the absence or even loss of less favored varieties. These less favored varieties may have many desirable and locally valuable attributes such as drought resistance and resistance to local pests.

When variety is suppressed, the dominant variety may become vulnerable to local conditions and require substantial chemical intervention in the form of increased water usage, fungicides, insecticides, and other toxic substances. Monoculturing of food supplies raises a host of concerns about the loss of locally viable and useful crops.

To respond to some of these concerns, organizations have started to bank seeds from a wide range of plants. A group of botanical gardens and museums has joined with private businesses to catalog and save as much of the cultivars of our plant population as possible in the Millennium Seed Bank Project. This project, managed by the Royal Botanic Gardens of Kew in Great Britain, is an example of ex situ preservation of species, preserving the species outside its naturally occurring habitat. This type of preservation has had a long, increasingly controversial, history as applied to zoo conservation of animal (and human) species. ***See also* Volume 2, Chapter 4: Biodiversity and Monoculture.**

Reference

Millennium Seed Bank Project online at www.kew.org/msbp/visit/index.htm.

Genetically Modified Organisms (GMOs)

The study of genes and associated sciences has developed to the point of permitting humans to create new species of plants and animals by combining their DNA in new ways. This kind of manipulation of plant and animal life has raised many ethical questions, popular controversy, and scientific questions about long-term consequences.

ECOSYSTEMS

Ecosystems are groups of natural processes that are connected through basic chemical, thermal, and biological interactions. In Western science, this term is limited to the physical components of a particular region such as the grouping of plants, animals, and other geographic features in a particular area. In other cultures, such a unit would also include intangible and metaphysical features such as the presences of beings that had lived and died in that place, the people who currently lived there, and ancestors.

The space-time qualities of these natural groupings are also described as biomes, ecotopes, and ecotones. These are the basic building blocks of sustainability from the environmental perspective. They are not neatly ordered, and the terms are used loosely. In policy areas, ecosystems are difficult to administer because they can cross local, national, and international boundaries. Most policy approaches to ecosystems examine the watershed and the flora and fauna, and the marine and terrestrial life forms contained in it. Successive and complete food chains, a range of biodiversity, and interdependence of natural systems indicate a healthy ecosystem. Intermittent and incomplete food chains, degraded biodiversity, and tattered interdependence of natural systems indicate an unhealthy ecosystem.

Ecosystem health can often be a matter of the length of time under consideration. An environmental perspective of sustainability defines the relevant cycle of time as that which is necessary for a healthy ecosystem, depending on climate and landscape of the place. Both industry and environmentalists want to know the carrying capacity of a given ecosystem. How much human impact can a given ecosystem take? Environmentally based sustainability advocates and others are curious about their own impact on their ecosystem.

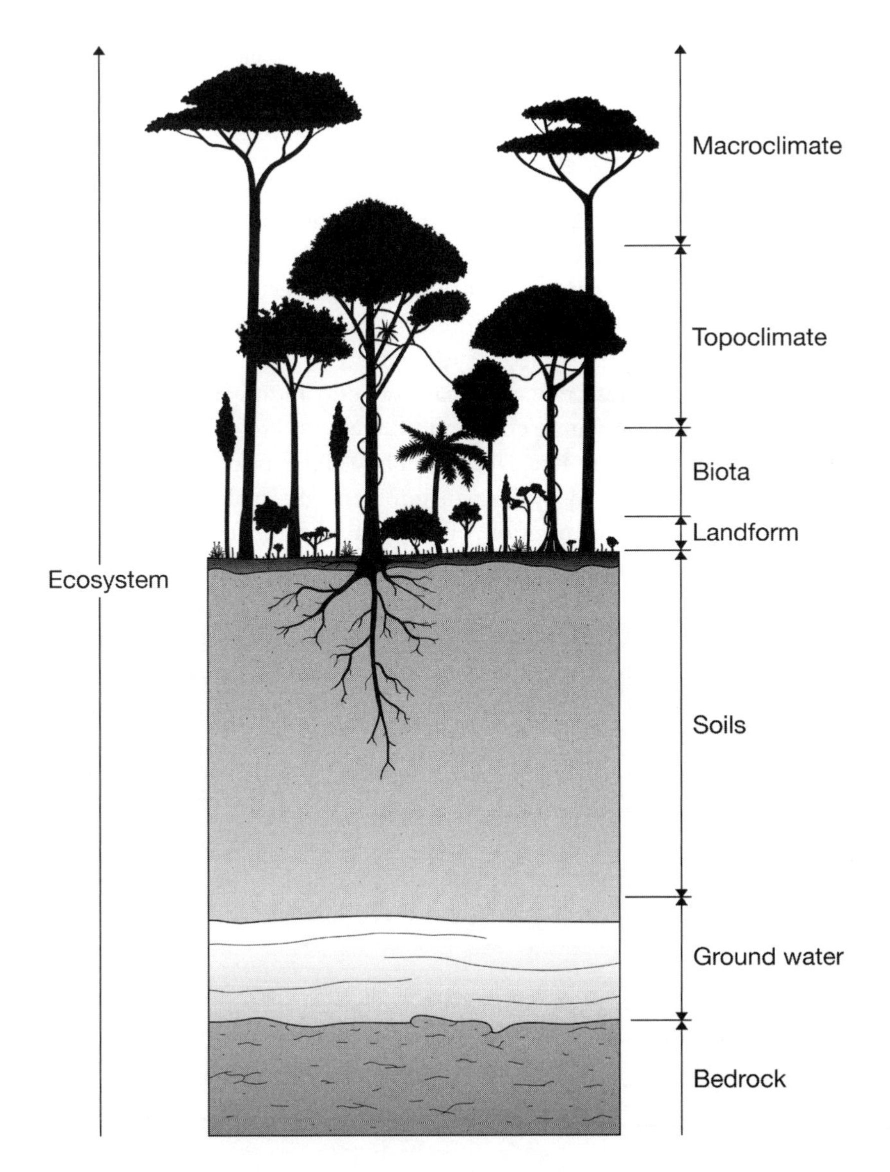

FIGURE 2.11 • An ecosystem describes the interaction between land, air, water, plants, animals, thermal, biological, and chemical dynamics in a given area. Illustrator: Jeff Dixon.

The U.S. Environmental Protection Agency's Council on Environmental Quality defined ecosystem as:

> An ecosystem is an interconnected community of living things, including humans, and the physical environment with which they interact. Ecosystem management is an approach to restoring and sustaining health ecosystems and their functions and values. It is based on a collaboratively developed version of desired future ecosystem conditions that integrates ecological, economic, and social factors affecting a management unit defined by ecological, not political, boundaries.
>
> Twenty-fourth Annual Report of the Council on Environmental Quality, 1993.

Reference

Klijn, Frans. 2007. *Ecosystem Classification for Environmental Management.* New York: Springer.

The Twenty-fourth Annual Report of the Council on Environmental Quality. 1993. Available on line at ceq.hss.doe.gov/nepa/reports.

Ecosystem Services

Ecosystem services come from ecosystem interactions and have uses that benefit humans. These benefits consist of uses that benefit humans that come from ecosystem interactions. Ecosystem services are benefits to humans that form the base of community and commodity. They include climate control, fresh water supplies, living soil, plant pollination, basic food production, most raw materials, genetic biodiversity, and medicines. It is difficult to put an economic value on these environmental resources because they are irreplaceable. They are resources that are finite, and are used and abused by human impacts. Sustainability proponents want to minimize consumption of essential natural resources by measuring and understanding ecosystems and ecological limitations. This requires natural resource conservation and enhancement, and a strong public policy regime around sustainable resource management of soil, water, air, agriculture, energy, and transportation. By valuing ecosystem services, sustainability advocates hope to push for cleaner production processes for goods and services and to greatly reduce waste.

The excessive use of ecosystem services is most prevalent in the developed nations of the world, where approximately 25 percent of the population consumes roughly 75 percent of the global resources. Ecologically based sustainability will be a big challenge for North Americans and may require major and unforeseen lifestyle changes.

References

Giudice, Fabio et al. 2006. *Product Design for the Environment: A Life Cycle Approach.* Boca Raton, FL: CRC Press.

Hunkeler, David et al. 2008. *Environmental Life Cycle Costing.* Boca Raton, FL: CRC Press.

Environment in International Application

The term *environment* can have different meanings when used in an international context. It does not always include ecosystem considerations. It can mean environmental security that can refer to adequate food and water supplies. It can mean environmental health that includes human rights. The line between environmental protection and human rights shifts in the international context. Environmental rights can include human rights in certain contexts such as forced relocation of people, effects on women and children, the rights of indigenous people, and effects on vulnerable populations. Vulnerable people include the young, sick, and elderly.

From an international transactional perspective on the environment, human rights are considered on a project level, not a national level. Environment can also mean the way environmental impacts are handled in large international financial transactions. In these contexts, the ecosystems are given greater consideration.

There is no single, unified international standard of practice or measurement for impacts or for monitoring of these impacts on the environment. There are international treaties and agreements, United Nations Standards, the World Bank's Pollution Prevention and Abatement Handbook (go.worldbank.org/ZBU119DC21), the World Health Organization suggestions (www.who.int/hpr/NPH/docs/hp_glossary_en.pdf), and the standards of individual countries. Some specific agreements that can be used are:

- The Malpol Convention (www.jbic.go.jp/en/faq/guideline/002/index.html)
- The World Heritage Convention (www.environment.gov.au/index.html)
- The Aarhus Convention (ec.europa.eu/environment/aarhus)
- The Ramsar Convention (www.ramsar.org)
- The Washington Treaty (www.nato.int/docu/basictxt/treaty)
- The Red List of the International Union of Concerned Nations (www.iucnredlist.org/static/organization)
- Directives regarding resettlement and indigenous peoples of the World Bank

There are many other international agreements, formal and informal, that circumscribe the term *environment*. As world concern about sustainable development increases, differences in the meaning of the term create confusion, controversy, and conflict as nations seek to increase the quality of their citizen's lives and other nations seek to decrease environmental impacts that irreparably destroy systems of nature on which all life depends.

All these agreements, treaties, and standards are difficult to enforce and monitor. Many use the term *environment* in ways that conflict with each other. One of the main differences is whether human rights issues, ecosystems assessments, and community involvement are considered. When large projects are developed that require international sources of revenue, however, some of these issues can be seen more clearly.

In large international financial transactions, *environment* is a term that is often controlled by the source of the financing. Most large financial lenders now want to prevent any adverse environmental impacts and to control them when they happen. They will either deny the loan, or if negative environmental issues arise, suspend loan payments. Some lenders go beyond that and want to achieve environmental improvements. By environmental improvements, they mean environmental conservation, consideration of alternative energy applications, and protection of the global environment. Protection of the global environment generally refers to greenhouse gas emissions, especially carbon dioxide. In terms of the application of the term *environment* to decision making, most large lenders profess to want these processes to be transparent to all stakeholders. This generally means that public notice of the decisions and the ability to comment on the decision are available to the public in the place of the environmental impacts.

Environmental assessment and monitoring are supported by international financing of big projects. These can be political and controversial. Nongovernmental organizations are now recognized by many international agreements and treaties and will often monitor and assess the environmental impacts of a project. Given the emphasis on transparency, it is now difficult to ignore these groups and they could derail a big project. Most international sources of funding want the project proponent to first review the environmental impacts on the project. They also require the project proponent to monitor the project for environmental impacts. An open question is whether they require the project proponent to mitigate the environmental impacts, and whether those environmental impacts will be monitored. Because the term *environment* can mean impacts on human rights, these dynamics can be very political. Most lenders require that the environmental assessment procedures of the host nation be followed and be made available to the impacted community. Some require consultations with the host community and that these consultations are part of the project plan.

A big question surrounding the term *environment* in international transactions is who the stakeholders are. Generally, they are the local community and local nongovernmental organizations. Sometimes they are all those who are at risk of being adversely affected by the project. Because the term *environment* can include human rights, those adversely affected can include a larger group of people than those who are on exposure vectors of emissions. They can include people in contiguous nations. This also heightens the political context, but may also expand the environmental impact assessment into broader ecosystem considerations. An open question is whether to consult with civil societies, communities, and nongovernmental organizations outside the host country.

In international projects with environmental impacts, there is great concern about the flexibility of the standards. Some of the issues raised about standards include whether they will take into account the vulnerability of the natural environment, how to deal with human corruption, whether unsustainable projects should progress contrary to community monitoring, whether standards include monitoring of ecosystem impacts on *potentially* endangered species, and who does the monitoring and reporting.

These are difficult and unavoidable questions engendered by the ambiguity of the term *environment* on an international level. There is concern from environmentalists that these ambiguities can be used to exploit areas that are not yet aware of their levels of ecosystem capacities. In areas with little or no human habitation of large project development, it is unlikely that the term *environment* is used. Sustainable development, however, will require knowledge of these areas to determine their level of vulnerability and carrying capacity. ***See also* Volume 2, Chapter 2: Language of Sustainable Business.**

References

Dietz, Thomas, and Paul C. Stern. 2008. *Public Participation in Environmental Assessment and Decision Making.* Washington, DC: National Academies Press.

Gibson, Robert B. et al. 2005. *Sustainability Assessment: Criteria and Process.* London, UK: Earthscan.

Kassim, Tarek A., and Kenneth J. Williamson. 2005. *Environmental Impact Assessment of Recycled Wastes on Surface and Ground Waters.* New York: Springer.

Schaltegger, Stefan et al. 2006. *Sustainability Accounting and Reporting.* New York: Springer.

Ecological Footprint

To assess the effectiveness of these strategies, measures and baselines of environmental impacts must be established. The amount of resources, including energy, needed to support specific activities can be calculated. This amount reflects the imprint of that activity on the Earth's natural resources. There are many ways to think about environmental consequences and how to measure them. Some calculate only greenhouse gas-creating activities. Others look at additional types of impacts on surrounding land and water resources. Individual calculators let individual people measure their impacts on their environment by looking at their daily life activities and lifestyles. Calculations for businesses and organizations tend to be more complex. They can be based on emissions measurements or mass balance methods. Mass balances look at information about materials and energy entering a production system

Ecofootprint Analysis

One of the most comprehensive online calculators for individuals is sponsored by Redefining Progress, an organization of economists. Another similarly comprehensive calculator is hosted by the Earth Day Network. You can take their tests online at www.myfootprint.org/en/redefining_progress/about_us/ or www.earthday.net/footprint/index.html.

Several different calculators for companies can also be found online. Business calculators may require a business to make several assessments. Some calculators focus on greenhouse gas emissions from commuting or business-related travel. Some focus on electricity usage. Global Footprint Network has footprint calculators for individuals, businesses, cities, and nations. See www.footprintnetwork.org/en/index.php/GFN/.

Colleges and universities have individually created audits that combine information from multiple sources.

and calculate emissions and materials left at the end of the physical cycle. Some important environmental regulations rely on businesses to do their own measuring and reporting of their environmental impacts that they are required to submit to the government on their oath that it is true.

FINANCE

The modern financial world is characterized by participation in the marketplace for debt and investment, not simply safekeeping of currency or valuables. Debtors are seeking cash for consumer and household purposes. Businesses seek cash for purposes of conducting their businesses. Nations participate in this market as well when a nation seeks cash to support its national projects like the development and renewal of its public infrastructure, or war. In that marketplace, promises to pay for loans of money between individual borrowers and lenders are reflected in credit cards and loans. When the borrower is a corporation or a nation, its promise to pay is called a bond. All such loans are made in exchange for a promise to pay additional sums of money as interest or fees. In some cultures and countries that prohibit the charging of interest, other types of accommodations may be made in exchange for the loan of money for investment.

In addition to bonds, individuals, organizations, and public governments may purchase part ownership in a business called shares. This ownership entitles the shareholder to participate in the profits made by the business. If a government has granted limited liability to a corporation, then its shareholders will not have to bear the risk of losses if the company incurs them, at least not beyond the value of their shares.

In this sense, all lenders and shareholders participate in the economic venture of the borrower. If the borrower does well, the lender and investor will also make money. In this sense, financial markets are an expression of confidence in the goals, mission, and abilities of a company. If the only motivation for investment or participation is financial return in the form of cash payments, sustainability as a goal may not mean anything at all to investors and shareholders. In a global capital market, those shareholders and investments may disinvest in a business almost instantaneously based on short-term gains and losses. This can discourage businesses that want to change their business production or methods in ways that reap long-term benefits but near-term losses.

Sustainability is about future generations, and, in that sense, it is about the long-term view of economic uses measured against the effect on the natural world. Short-term investors may desert companies pursuing sustainability for short periods. Businesses with a longer term view, however, will expect to regain financial investment costs over time. The perspective of time is an important element of financial confidence and communication. To make sustainability a goal of investment, short-term losses must be discounted in favor of future generations and future viability.

Some groups of investors have begun to demand accountability of their borrowers and lending projects. They have asked for greater transparency and accountability for the use of their capital funds in terms of environmental and equitable impacts. Some require this accountability at the outset of a project in terms of planning. Others require it on an on-going basis in the form of annual reports and annual accounting. The trend is growing to require both front-end and regular accounting for the social and economic consequences of business operations.

When nations involved in development projects are the borrowers or lenders, special terms may be used to describe their roles as lenders. These terms may make it difficult to understand the nature of their participation in the usual transactions of lending, but they are different from other types of lenders only in terms of scale, not a difference of kind.

Development Banking

To develop and grow, nations and businesses often require large investments of money to build basic national infrastructure like roads, ports, dams, and electric plants. This infrastructure is essential to allow businesses to transact with others. Loans of capital are also used to purchase technology. Technology is as important as these other types of traditional infrastructure in the contemporary age.

Sources of money for development can come from many sources. Governments give money to other governments in the form of official development aid (ODA). Other sources of development funding include individuals and groups of private investors such as institutional investors, public investors, and quasi-public organizations using combinations of funding from both public and private sources. ODA is a far smaller source of capital for development than private and quasi-private sources.

Some development banks were created to distribute funds for these purposes from several nations. These are called multilateral development banks. Other development banks lend money from private sources for these purposes.

Development banks specialize in some types of projects associated with heavy ecological footprints such as mining, oil and gas extraction, paper and pulp, and agribusiness. They channel funds from various sources into projects for development. They operate on the expectation of profit and charge fees and interest for the loan of their capital. Many development banks are able to take a longer term perspective on the need for profitable return because of their unique mission.

Development banks use social, political, and environmental risk assessments to evaluate investment projects. These banks may make demands of their nation-state borrowers that reflect their interest in long-term profitability. These demands, called structural adjustments, can include requirements for adjustments and restructuring of national priorities that would be unusually intrusive in the ordinary loan between individuals and a bank. For example, some of these structural adjustments could include national decisions about social matters, not usually considered in a business relationship between lender and borrower. Some of the structural adjustments have included demands for legislative reform of land ownership, agricultural practices that may damage the environment, and cuts in education and public and social safety nets. ***See also* Volume 2, Chapter 4: Development Lending.**

The World Bank

The World Bank is a multilateral development bank. Initially, its funds came from the nations who won World War II and decided to rebuild their former enemies. Now, its funds are primarily used to promote development of underdeveloped nations. The World Bank became embroiled in controversy for its structural adjustment requirements in developing nations. It led the way in developing environmental and social principles of investment, later adopted in the Equator Principles that govern its projects in developing countries. ***See also* Appendix B: The Equator Principles.**

Investment Promotion Agencies

Governments form agencies especially for the purpose of encouraging exports and other investments. To do this, they offer credit financing, risk insurance including political risk insurance, guarantees, and other incentives to facilitate investments. In addition, promotional agencies provide insurance and other assistance to facilitate export and investment. Promotional agencies may be funded by governments or private parties.

THE EXPORT-IMPORT BANKING

An export-import Bank is a promotional agency of the government. It specializes in providing credit insurance and guarantees to facilitate the purchase of U.S. goods by foreign nations.

THE BERNE UNION

The Berne Union is the largest association of development banks, promotional agencies, and private insurers in the world. It manages the largest amount of money for development lending in the world, far exceeding direct development aid. In their guiding principles, Berne members state, "We are sensitive about environmental issues and take such issues into account in the conduct of our business." They also state, "We are sensitive about environmental issues and take such issues into account in the conduct of our business."

References

Freidman, Thomas L. 2000. *The Lexus and the Olive Tree: Understanding Globalization.* New York: Anchor Books.

Gray, John. 1999. *False Dawn: The Delusions of Global Capitalism.* New York: New Press.

Risk Assessment and Development Banking

Risk assessment is used in decisions about development projects, and it is used by governments in budgets and planning activities. Risk assessment analyzes the probability and magnitude of harm from various events and activities. It is widely used to make decisions in development finance. Insurers rely on this computation of risk of harm in making decisions about whether to insure, and if so, how much to charge for insurance. Related to the science of risk assessment, risk management determines how to plan for and communicate about risks. Risk perception is a science devoted to examining the qualitative aspects of risk, not simply its quantitative aspects.

Risk science posits that danger is real, and dangerous conditions may be present in a wide variety of circumstances from weather hazards to human policy-oriented decisions. Risk scientists try to identify what dangers exist across a broad spectrum of environmental and human circumstances, and the probability that these dangers will affect a project. They also calculate the dollar value of damage when and if such dangers should actually occur. These calculations go into the decision about whether, when, where, and how much to invest in projects. They can also weigh into decisions to require conditions such as structural adjustments for project investment.

Both the Equator Principles and the Berne Union specifically rely on risk assessment as a matter of principled decision making about their loan projects. The Equator Principles state

> **Principle 2: Social and Environmental Assessment**
>
> For each project assessed as being either Category A or Category B, the borrower has conducted a Social and Environmental Assessment ("Assessment") process to address . . . the relevant social and environmental impacts and risks of the proposed project (which may include, if relevant, the illustrative list of issues as found in Exhibit II).

The Berne Union states in its principles only that "We carefully review and manage the risks we undertake." *See also* **Volume 2, Chapter 4: Development Lending, Appendix B: The Equator Principles.**

Millennium Development Goals

In 2001, in recognition of the new millennium, all 189 United Nations member states adopted the United Nations Millennium Declaration. This declaration laid out eight goals for human society and government for the next millennium:

1. Goal 1: Eradicate extreme poverty and hunger
2. Goal 2: Achieve universal primary education
3. Goal 3: Promote gender equality and empower women
4. Goal 4: Reduce child mortality
5. Goal 5: Improve maternal health
6. Goal 6: Combat HIV/AIDS, malaria, and other diseases
7. Goal 7: Ensure environmental sustainability
8. Goal 8: Develop a global partnership for development

Progress has been slow and frustrated by the failure of major developed nations like the United States to deliver aid in the amounts promised, and by escalating environmental degradation.

Beyond political instability, some economists and businesses perceive a business opportunity in empowering poor people and poor communities. Mohammed Yunus, banker and professor, and winner of the 2006 Nobel Peace Prize for his work in poverty reduction, founded the Grameen Bank, and created a financial sector called microfinance that has opened credit opportunities for people without the traditional criteria of creditworthiness.

Mohammed Yunus

Mohammed Yunus is known as the "banker to the poor," also the title of his autobiography. Yunus founded the Grameen Bank in Bangladesh in 1983 after teaching economics at the Chittagong University. While a professor at the university, he took his students on a field trip to a poor village. He spoke with a woman there who was a basket weaver. She told him about her plight—that she needed to borrow money to buy the bamboo for her baskets, and the lender charged weekly interest that took all but a penny profit margin. She was working but unable to pull herself or her family out of abject poverty. This experience changed Yunus's mind forever. He started to lend small amounts to people personally, mostly to poor women who were in situations similar to that of the woman he met in the village that day. His experience with these small personal loans convinced him that poor people had the desire and ability to participate in businesses to lift themselves from poverty, but they were denied access to capital because there was no way to value their qualities in traditional banking terms.

Against all the advice from banks and friends, he started the Grameen Bank, which focused on microlending as a means to eradicate poverty. Today, this bank has a loan repayment record of 98 percent; more than 94 percent of borrowers are women. The bank has grown to serve more than 2.1 million borrowers and receives $1.5 million a day in weekly installments.

Yunus and Grameen Bank won the Nobel Peace Prize in 2006. In accepting the prize, Yunus said, "Lasting peace cannot be achieved unless large population groups find ways in which to break out of poverty. Micro-credit is one such means."

Reference

Grandin, Karl, ed. *Les Prix Nobel. The Nobel Prizes.* 2006. [Nobel Foundation], Stockholm, 2007 nobelprize.org/nobel_prizes/peace/laureates/2006/yunus-bio.html, muhammadyunus.org/content/view/47/69/lang,en/.

Asset Building Movement

The traditional indicators of economic stability are a job and an asset. These are the fundamental signposts that an individual or family is on the way to a middle class standard of living. Beyond that, wealth indicators include savings, retirement benefits, health benefits, and educational opportunities. Poor individuals and families often do not have any of these indicators. The asset building movement is committed to transforming public policy benefits into assets convertible to these more traditional assets. Sometimes this can take the form of government committing to match small savings accounts put aside by poor families, usually on a 3:1 basis. Under this type of plan, saving $42 a month can turn into $6,000 in three years. Other forms of asset building involve allowing poor people and families to draw on a lump sum set aside for crises like unemployment for the purpose of buying a home or paying for education.

GLOBALIZATION

Markets were local events until the growth of interconnected communications. The ability to transact business by telephone, facsimile machine, and computers connected by the Internet has meant that market transactions are no longer tied to a specific location and time as they once were. Markets are now virtually ubiquitous. Buying and selling of goods and services transcends national boundaries and laws and can occur at any time and in any place that is connected to the Internet web of communications. This has made markets global in scale.

These global markets depend on an infrastructure of wires, cables, and other wireless technology to communicate and transfer intellectual property and financial commitments around the globe almost instantaneously. To the extent that these markets exchange goods, they rely on the more traditional infrastructures of roads, railways, shipping lanes, and ports for the physical carriage of goods to deliver products throughout the world. To the extent that fossil fuels are necessary for the physical transfers of goods, globalization of products and their manufacture may be slowed as the cost of transport exceeds the cost of labor.

The growth in markets has greatly increased growth in businesses, both manufacturing and financial sectors. Market growth has also increased the environmental footprint of businesses on the ecosystems of our planet, and especially on the dynamic of climate change. If markets continue to grow as they have, the climate change and ecosystem impact of that growth will tip the planet into ecosystem chaos. Sustainable development contemplates continued growth and development that is based on energy and resource uses that do not damage the planet, or that could even help to restore the planet's environmental systems.

Markets have been good vehicles for increasing wealth, but growth in marketplaces has also accompanied alarming trends in the rate of climate change and other types of environmental deterioration, as well as growth in poverty. This has been established by the work of the Millennium Ecosystem Assessment (MA). Its findings outlined the negative impact of human actions on the Earth's natural capital. The MA showed that humans cannot take for granted the Earth's ability to sustain future generations at this rate of environmental degradation; however, if appropriate substantial action were to be taken, it would be possible to reverse some degradation over the next 50 years.

Environmentally unstable or deteriorated places also are likely to be politically unstable. Without peace, economic growth and stability are simply impossible because sellers and manufacturers cannot be sure of their ability to grow crops, manufacture goods, or deliver those goods through the physical obstacles that confront them. In addition, investors will not risk investing capital in unstable regions.

Although private market-based connections have spread across the globe, governments have also facilitated the trade in goods and services across geographical boundaries by agreeing to remove customary barriers to trade like importation taxes (often called customs duties). The largest of these agreements is a multilateral agreement called the World Trade Organization (WTO). There are also regional agreements of this type like the North American Free Trade Agreement (NAFTA). These agreements can be controversial within member nations. Some argue

WTO structure

All WTO members may participate in all councils, committees, etc, except Appellate Body, Dispute Settlement panels, Textiles Monitoring Body, and plurilateral committees.

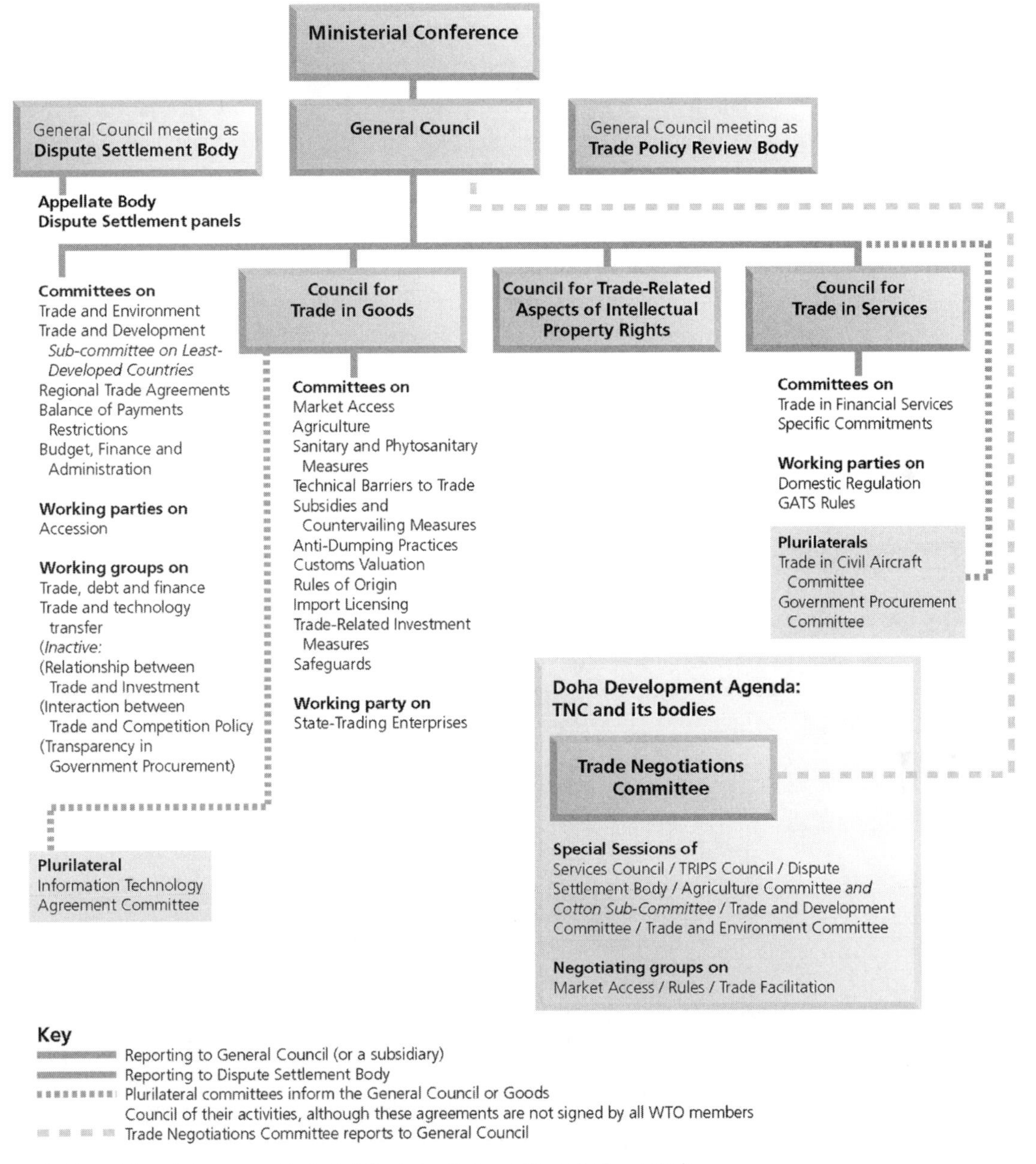

FIGURE 2.12 • The structure of the World Trade Organization. Courtesy of the World Trade Organization.

that these agreements diminish environmental and labor protections in order to compete based on price with countries that have lower standards of protection. *See also* **Volume 2, Chapter 4: Localization; Volume 2, Chapter 5: Globalization and Localization.**

KEY BUSINESS SECTORS

For many purposes, businesses are classified and grouped together by common characteristics. Two systems of industrial classification are the Standard Industrial Classifications, and North American Industry Classification System (NAICS). The United Nations has its own classification system, the International Standard Industrial Classification System (ISIC). Businesses grouped together form a sector for some governmental or analytical purposes. In terms of environmental analysis, some industrial sectors have a much larger environmental footprint and trajectory than others.

The Environmental Protection Agency (EPA) used this classification system to identify sectors with the greatest impact on the environment in terms of waste and pollution. The idea of using these classifications to identify and focus on those businesses affecting the environment most severely allows regulation and assistance such as clean production technology to be tailored to the requirements of an industrial sector and efficiently disseminated throughout that sector without creating unintended disincentives such as competitive disadvantage. Thirteen sectors are currently working on sector-based plans to diminish their impact on the environment through environmental management systems and other strategies:

- Agribusiness
- Cement Manufacturing
- Chemical Manufacturing
- Colleges and Universities
- Construction
- Forest Products
- Iron and Steel
- Metal Casting
- Metal Finishing
- Oil and Gas
- Paint and Coatings
- Ports
- Shipbuilding and Ship Repair

Extractive Industries: Timber, Grazing, Mining, Oil, and Gas

Industrial sectors that depend on the consumption and extraction of specific resources from the Earth's crust play a vital role in ecosystems and their balance. For economies that enjoy this natural advantage, the financial significance of these resources is enormous. Harvesting and selling these resources have often been the engine for development in resource-rich–cash-poor countries. These industries and the resources they extract are often at the center of environmental degradation, including loss of rainforests, loss of species habitat, and desecration of sacred sites of indigenous people. ***See also* Volume 2, Chapter 4: Unsustainable Industrial Sectors.**

Agriculture and Sustainability

Agriculture is what allows human populations to grow and thrive. Societies that have transcended the physical limits of hunting and gathering on a day-to-day basis for their sustenance have achieved some level of proficiency at growing and storing their food through the practice of agriculture. Although agriculture is practiced in a broad variety of forms, governmental policies have successfully maximized certain forms of agriculture. In many countries, farmers are still primarily women tending small plots of land, often at considerable distances. In a globalized economy, however, most food traded in the global marketplace is produced by agribusiness, a corporate industrialized form of agriculture.

As agribusiness grows in scale, it often eliminates the rural communities around it. Small farms will be purchased or leased to increase field acreage, moving family-based farming operations out of the area. Small businesses that served smaller farmers also become redundant as large-scale operations provide their own in-house services such as slaughtering, butchering, and machinery maintenance.

Agribusiness is a commitment to agriculture of scale, achieving economies of scale, and externalizing costs of production such as water pollution through runoff, and loss of wetlands and carbon neutralizing forests through drainage programs. Sometimes, a similar scale of labor is needed for periodic harvesting and plantings, although heavy equipment has often replaced human labor in these operations. As the price of fuel increases, the cost of operating heavy equipment may outstrip the cost of labor, especially if products are being grown to be transported long distances. Early in the industrial development period, slavery was promoted as a public policy designed to encourage agribusiness in labor-intensive areas like cotton and tobacco production. As slavery was outlawed, former slaves often stayed on the land as sharecroppers, indentured servants, and later farm workers.

The Environmental Footprint of Agribusinesses

Agribusinesses tend to grow commodity crops for sale in the international marketplace. The crops they favor must pack and travel long distances well. This leads to the mass production of these types of crops, resulting in the disuse and noncultivation of other types of plants that may be more suited to small, regional conditions. Agribusiness has made production of livestock an indoor activity rather than a field-based activity. When large herds of animals like chickens, hogs, and cattle are kept in confined conditions, they create vast lagoons of fecal waste that contribute significant pollution to local lakes, rivers, and streams.

Agribusiness farms large tracts of land using heavy machinery, often clearing away forests to plant plantation-style crops. The practice of clearing and tilling huge tracts of land can have a devastating effect on topsoil as it dries out and blows away.

The environmental footprint of this type of farming is felt globally through climate change and desertification on vast scales. In addition, governments have often undertaken the cost of constructing infrastructures that support large-scale agriculture such as massive irrigation projects bringing water to dry places and drainage projects that make wetland farmable. Over time, however, these massive kinds of projects create negative environmental effects themselves, causing salinization of the soils as water runoff increases, carrying chemicals and minerals from fertilizers and pesticides into the soil. Drainage of wetlands also increases water pollution, as wetlands are one way in which nature cleans water. Finally, these projects also tend to attract development for residential and other commercial uses that will compete with agriculture for water and land on which to build. These uses also increase runoff and the downward spiral of environmental impacts takes hold.

Agricultural Subsidies and Sustainability

Many governments want to encourage production of food in order to ensure food security for their populations. These governments may use legal and policy tools like tax breaks and subsidies to encourage investment and development in this sector. When these policies reflect the goals of a different age—the age of industrial style development—tax breaks and subsidies flow to unsustainable crops and practices. Sometimes these governmental incentives encourage the cutting of large forests for the sale of timber, without considering the global costs in terms of increasing levels of carbon dioxide in the atmosphere and acidification of the oceans. It is hard for developing nations to consider these factors when their people are hungry, children die from preventable diseases, and everyone needs access to education. The prospect of agribusiness providing income for workers and revenue to the economy is what motivates some contemporary unsustainable behaviors in poor and developing areas.

Subsidies help agribusiness grow in size and influence the choice of what it produces and how. Economies of scale make agribusinesses successful, and these are due not simply to their own internal economies, but to government subsidies paid to promote the growth of certain commodities, and sometimes to promote conservation. These subsidies are mostly received by agribusinesses, not individual farmers or farm families living on the land they work. They also make the products of one country cheaper on the international marketplace than the products of another country. That is one reason why agricultural subsidies are disfavored by international trade organizations like the WTO and NAFTA. Countries that continue to pursue subsidizing commodities traded internationally risk sanctions from both of these organizations. ***See also* Volume 2, Chapter 4: Agribusiness and Sustainability.**

Reference

Clapp, Jennifer, and Doris Fuchs, eds. 2009. *Corporate Power in Global Agrifood Governance.* New York: Columbia University Press.

Chemicals

This industrial sector is responsible for the production of the chemicals that have transformed modern life for good and for ill. Their products include fertilizers and pesticides so important in the food production businesses. These chemicals also pollute sources of freshwater on which many forms of life depend. Their products are also essential to many production methods, but can cause production methods to contribute to pollution as a by-product. Their products are also ubiquitous in consumer products from pharmaceutical drugs and cosmetics to paints and finishing products.

The number of manufactured chemicals that humans encounter in daily life soared after World War II because of scientific experimentation to produce weapons. Some estimate that there are 116,000 manufactures chemicals in regulated tap water. Chlorinated substances became widely available for consumer and household uses such as laundry and sanitization. The risks posed by the presence of these chemicals alone, in combinations, and synergistically have never been fully investigated.

The effects of these chemicals alone or in combination on human health, the health of other species, and ecosystems as a whole are contested. Some have connected chlorinated chemicals and other synthetics with cancers, reproductive disorders, and other health challenges.

Transportation

The movement of raw materials, labor, goods, and services affects business and is a business. It also affects energy use and other impacts on the systems of nature on which future life depends. Reliance on nonrenewable and dirty energy sources such as gas and oil creates large global impacts. The costs and impacts of transportation are one set of reasons why cities are arguably more sustainable than sprawling suburbs. It is also a business that is directly and indirectly subsided by governments seeking economic development. The idea of sustainable transportation is to allow the free movement of people and materials without harm to the environment or public health.

As a business sector, transportation undergoes tremendous technological entrepreneurship in alternative energy, infrastructure design, and environmental impact assessment. Transportation is also implicated in disproportionate environmental impacts by race and class. Transportation public policy decisions are often large and difficult to undo. As such, they often involve major political and economic controversies. The construction of large transportation systems often has environmental impacts themselves.

References

Bullard, Robert D., and Glenn S. Johnson, eds. 1997. *Just Transportation: Dismantling Race and Class Barriers in Mobility.* Gabriola Island, BC: New Society Publishers.

Davenport, John, and Julia L. Davenport. 2006. *The Ecology of Transportation: Managing Mobility for the Environment.* Boca Raton, FL: Springer.

Janic, Milan. 2007. *The Sustainability of Air Transport: A Quantitative Analysis and Assessment.* Surrey, UK: Ashgate Publishing.

Schafer, Andreas et al. 2009. *Transportation in a Climate Constrained World.* Cambridge, MA: MIT Press.

Tolley, Rodney. 2003. *Sustainable Transport: Planning for Walking and Cycling in Urban Environments.* Boca Raton, FL: CRC Press.

LAND, SOIL, AND FORESTS

Land is the soil and products of the soil. Minerals, subsurface waters, and vegetation can be viewed as part of the land. About 25 percent of Earth's surface is land. Direct human habitation is on approximately 1 to 2 percent of the land, although the impact of human habitation is far beyond that. Land can often be reduced to a parcel of land with definite boundaries up, down, and around it. These boundaries can be human-imposed political and real estate boundaries or natural boundaries like watersheds. Human-imposed boundaries can be linked to resource extraction and to ecosystem serves like agriculture, mining, or fresh water. These boundaries can also be linked to urbanizing actions like real estate development and the built environment of houses, offices, shopping malls, and factories. Land is also an economic concept that forms the basis of private property. In this context, it is a bundle of legal rights, responsibilities, and obligations. It has economic value fundamental to capitalism and to liberty concepts. For many indigenous people, land conveys a meaning through a sense of place that gives them a cultural identity.

Most of the Earth is covered by water, mainly in the form of salt water. Land is the dry surface portion of Earth's crust. The area of transition from land to water is constantly changing. See Wetlands and Estuaries defined in Volume 1, Chapter 2.

Dry land itself has many levels, including the topsoil that is essential to human agricultural cultivation and grazing, as well as deeper levels that contain minerals, oil, and gas. Water resources are also found within these deeper levels. See discussion of groundwater Volume 1, Chapter 2. ***See also* Volume 2, Chapter 4: Private Property; Volume 2, Chapter 4: Real Estate.**

References

Amacher, Gregory et al. 2009. *The Economics of Forest Resources.* Cambridge, MA: MIT Press.

Bachmann, Peter, Michael Kohl, and Risto Paivinen. 1998. *Assessment of Biodiversity for Improved Forest Planning.* New York: Springer.

Moran, Emilio F., and Elinor Ostrom, eds. 2005. *Seeing the Forest and the Trees: Human-Environment Interactions in Forest Ecosystems.* Cambridge, MA: MIT Press.

Zarin, Daniel. 2004. *Working Forests in the Neotropics: Conservation through Sustainable Management?* New York: Columbia University Press.

Urban Sprawl

Sprawl is a contemporary land use problem linked to development patterns. A problem associated with sprawl is that people move out into natural areas, causing greater and greater impact on the environment and greater consumption of nonrenewable resources. Three factors fuel sprawl. First, the tendency of people to reside away from work requires transportation back and forth to work and home. Roads and motor vehicles greatly impact the environment in nonsustainable ways. Second, methods of commuting are inefficient, with many one-driver car trips. This also increases environmental impacts. Third, privacy and status values are reflected in a house, which tends to result in large-lot, low-density housing. This has a large environmental impact. ***See also* Volume 2, Chapter 4: Urban Sprawl.**

Brownfields

Brownfield land or brownfields are abandoned plots of land that were once used for commercial or industrial facilities. Some brownfield sites are the result of decommissioned military or industrial sites. This past use often resulted in contamination by hazardous waste or pollutants. Brownfields are specific plots of land that may be reusable once the land has been sufficiently cleaned up. Most brownfield sites have been left unused for significant periods; however, as land in certain locations becomes more scarce or expensive, brownfield sites become more valuable. Eventually they will be valuable enough to clean up to safe standards and redevelop. Furthermore, with increased precision and new techniques, the ability to bring brownfield land up to safe standards has become scientifically and economically feasible.

The regulation and cleanup of brownfield land is regulated by the EPA. The EPA works with individual states to provide technical assistance and cleanup of brownfields, as well as helping to determine sources of funding to ensure that brownfields are given new life.

Brownfield redevelopment is still not perfect, and some projects are abandoned because of rising costs resulting from unknown contaminants that exceed the initial evaluation. Most brownfield cleanup projects are for commercial use; however, there are some projects underway to determine whether brownfields can be used to grow crops. The intent is twofold, first to help with the cleanup process of the soil, and second to contribute a more efficient production of biofuels.

Greenfield land is undeveloped land currently being used for agriculture or undeveloped. Some greenfields are greenbelts that have prohibitions against development. Greenbelts are designed to protect the unique character of undeveloped land within areas of extensive development.

Greyfield land is land that was once thought to be economically profitable but eventually became obsolete and outdated. The term is usually applied to areas that were once considered viable retail and commercial plots of land that suffered from lack of reinvestment. Greyfields are not contaminated; instead, they are usually abandoned because of larger developments nearby. Greyfields may have a dormant value because they are often equipped with an underlying infrastructure, such as plumbing and sewage systems, that may be used if the land were to be redeveloped. Greyfields are those plots of land that have become stalled in their industrial development. If the area were to be rejuvenated, greyfields would be ideal places to resume commercial and retail investment because much of the basic groundwork may have already been laid. ***See also*** **Volume 2, Chapter 4: Urban Sprawl**

Reference

Christensen, Julia. 2008. *Big Box Reuse.* Cambridge, MA: MIT Press.

De Sousa, Christopher. 2008. *Brownfields Redevelopment and Quest for Sustainability.* Oxford, UK: Elsevier Science.

Dixon, Tom et al, eds. 2007. *Sustainable Brownfields Regeneration: Livable Places from Problem Spaces.* Hoboken, NJ: Wiley-Blackwell.

U.S. EPA Brownfields and Land Revitalization. www.epa.gov/swerosps/bf/index.html.

MARKET-BASED STRATEGIES

Human involvement with the marketplace has a long history. When ancient humans began to have surpluses beyond what they needed for themselves and their families, they would trade with others for other goods (called bartering), and eventually for money. That basic dynamic is the foundation of the global economy today. At first, markets developed by following animal trails and rivers that were later developed into the infrastructure of roads, railways, docks, and ports. These fundamental structures supported the growth of cities and towns and still support urbanization. Markets were initially sporadic and seasonal, but they have become a 24-hour/seven-day phenomenon in many countries, and everywhere that the Internet is available.

Trade both shapes and reflects the values of the traders. It reflects their tastes and convictions about what is necessary and beautiful. It also shapes what is produced in order to meet the demand of consumers. As consumers become more personally conscious about the consequences of their purchasing choices on our environment and on our communities, production will have an economic incentive to change to satisfy those additional criteria. This is sometimes referred to as the economic principle of supply and demand.

In achieving progress toward a sustainable economy, this dynamic illustrates the different but interconnected role that consumers and producers play. Consumers may influence the choices businesses make in terms of resources used, materials used, and labor conditions, if they want to do so. The larger the financial influence of a purchaser, the greater the influence. Consumers can make a market for a sustainably produced product by their individual choice. For example, the Department of the Navy will require all future facilities to use sustainable design principles, including energy conservation, reduction or elimination of waste and pollution, water conservation, and lifecycle management of resources and products used.

Another example of an influential individual purchaser is Wal-Mart's decision to purchase LED lighting for use in its stores. Wal-Mart is the world's largest retailer. It owns Sam's Clubs and Wal-Mart stores in 15 countries. It is the largest private user of electricity in the United States. To cut its energy usage, it has decided to use LED lighting in its refrigerated cases. This lighting is expensive initially, but pays for itself in energy savings. When a buyer the size of Wal-Mart or the Department of the Navy demands sustainability in production methods and products, the costs of production may well decrease as a result of economies of scale. This benefits all consumers regardless of size.

Market-based strategies to achieve sustainability use conventional trading incentives to achieve goals related to sustainability.

Critics of globalization and capitalism in general point to the fact that increased market presence has rapidly degraded our planet's environmental condition and added to a huge gap between rich and poor, even in developed nations. They question is whether anything inherent in the fundamental dynamic of trade can counteract these consequences. Others argue that market dynamics are tools that respond to the values of those who use them. They say that nothing about markets is inherently harmful to the environment or our communities. The ideas that they propose use the dynamics of buying and selling, but try to link them to the values of sustainability through legislation and regulation, or through voluntary business commitments.

Pollution Credits Trading: Cap and Trade Programs

Government can establish standards and limits on the ability of any company to pollute. If a company fails to meet these standards, it is subject to a penalty. If the company exceeds compliance, the government might choose to reward this achievement as an incentive. One way of rewarding the achievement is to issue a credit in the form of permission to pollute. By making such credits transferable, a market could be created around excess pollution.

Such solutions often fail to consider the consequences of market-based solutions on groups that do not participate in the market, such as poor people and their communities, and the ecological systems on which we all depend. When these groups are not considered, market solutions often fail to account for the negative consequences to those groups, resulting in replicating the downward spiral of human-driven harms to the environment and widespread poverty. Supporters of cap and trade programs point out that these programs can control and ratchet down all emissions so that that overall sustainable development is easier to achieve.

Environmental Consequences

These trading programs are being undertaken or proposed in the face of troubling factual ignorance in several areas: What is the actual amount of emissions currently affecting a region? What is a given ecosystem's carrying capacity? Where will pollution be transferred and what are the conditions at that new site? Permits are not required for every emission of pollution or waste into the environment. People and businesses are allowed by law to emit up to a certain threshold amount without the need to get permission. The amounts of these thresholds may be relatively small individually, but they can rapidly add up to a great deal of unaccounted for emissions taken together. We do not know how much, however, because this unregulated pollution is not reported. Nevertheless, indicators of ecosystem health will register these effects on the webs of life. Currently, there is no consistent effort being made to assess the carrying capacities of our ecosystems. There is no baseline and no monitoring of what is contributing to the load on our ecosystems and their services. Nor do such systems control for the impact of so-called excess pollution in the ecosystem and community to which it will be sent.

These uncertainties make it hard to say that trading pollution will result in less pollution and waste, if that is the goal of such programs. Capping and reducing such permitted emissions, however, is a key part of the argument in favor of these programs. The problem is that permitted amounts are rarely reduced, and the mere threat of such action has prompted stall resistance from permit holders. Assuming the political will exists to reduce permit levels, we are still faced with vast factual uncertainty about how much to reduce permitted emissions and how fast to do so in the face of no data about prethreshold levels of emissions and carrying capacity.

Consequences to Communities

Even if the environmental data were available and could be acted on effectively, these programs have not chosen to embrace responsibility for the ultimate destination of such excess pollution, most likely environmental justice communities or developing countries. When pollution is redistributed rather than reduced or eliminated, the mechanism reveals an underlying indifference about consequences that is unsustainable because of the interconnected nature of our ecosystems and communities. NIMBY is an acronym for the popular sentiment about pollution—"not in my backyard."

The challenge that such solutions must face is how to incorporate environmental and social consequences that are not salient in current marketplace terms. ***See also* Volume 2, Chapter 3: Role of Government in Moving the Economy toward Sustainability.**

Green Taxes, Carbon Taxes, and Green Fees

Government often tries to control market behavior through a system of subsidies and taxes. Taxes require payments from individuals or businesses to government. Rational tax policy generally adds such costs to things we do not want to encourage. Because most businesses and individuals do not want to pay taxes, they serve as a disincentive for conduct that government wants to reduce or eliminate. If government chooses to tax a particular product or act, that tax will be added to the costs of the activity or product. Adding such costs makes them more expensive and less desirable. Subsidies are government payments that reward conduct deemed desirable. Subsidies are usually added to things we do want to encourage. Tax policy for sustainability would add costs to the unsustainable aspects of goods and production methods and subsidize their sustainable aspects. This would encourage businesses to reduce or eliminate the unsustainable aspects of their operation.

Taxes on hydrocarbon produced by businesses could be used to discourage dependence on fossil fuels. This together with subsidies or tax credits for renewable fuels could shift our economy away from its use of unsustainable energy resources.

In 1993, Vice President Al Gore proposed a tax on users of fossil fuels roughly based on the greenhouse gas-generating properties of the fuel. The consequences of that approach to using the market to reduce fossil fuel use and force change to other energy sources would have been felt most intensely by poor people, people on fixed incomes, and households and individuals. This distributional effect would be considered an externality of a policy designed to make polluters pay for the use of fossil fuels. By not considering this type of externality, policies fail to incorporate the equity component of sustainability. Using tax revenues to offset these externalities is one way to approach fairness in pollution-based taxes. This is the idea behind Pigouvian taxes, named after English economist Arthur Pigou.

One type of problem with both green taxes and fees that can occur is taxing behaviors that are desirable in order to raise revenue for other sustainable behaviors. Similarly, subsidies policies become unsustainable when they are used to subsidize unsustainable behaviors.

Green Fees

Fees are costs added to the cost of a product or service. Fees can be added to discourage unsustainable behaviors in the same way that taxes are assessed. Revenues earned from these additional fees can be used to subsidize hardships falling unfairly on poor people or communities. ***See also*** **Volume 2, Chapter 3: Role of Government in Moving the Economy toward Sustainability.**

CLEAN PRODUCTION TECHNOLOGIES

Waste of resources and energy and pollution deposited into our communities and ecosystems from the production of goods and services are harming the environment and indirectly contributing to the problems of poverty. Excess, waste, and pollution stress the environment and leave poor communities unable to meet their basic needs. Eliminating waste and pollution would restore our environment and free more resources for poor communities and nations.

Engineers and architects, and artists and visionaries, have been engaged in multiple, creative experiments on how to redesign our production cycles so that they eliminate waste of resources and energy, as well as pollution and harmful ingredients from products. A few general principles characterize many of these approaches, including biomimicry, designing in ways that use more human labor and energy and less material. ***See also*** **Volume 2, Chapter 2: Language of Sustainable Business.**

Green Consuming

Market preferences expressed by consumers in terms of their purchasing habits can shape the behaviors and choices of producers. Consumers who are willing to purchase products with less impact on the environment, even if those products are more expensive than competing products, can influence producer choices toward clean production and more sustainably produced products.

Individuals who want to act based on their environmental and equitable ethics may be frustrated when confronted by their involuntary participation in marketplace transactions that harm their environments or communities. They have sought out opportunities to use purchasing power to change the material choices of their daily lives. Many businesses have focused on this new marketing potential by making changes to the production processes, their products, and their corporate policies. Even secondary business has thrived on being able to connect these types of consciously sustainable consumers and businesses.

Businesses are also consumers of goods and service. Through their relationships with suppliers and contractors, they may have great influence on other businesses in an economy by specifying the characteristics of what they want to purchase. Both business and government make consumer choices about materials and processes in which to invest, and their choices can have a ripple effect through an economy. When governments choose to purchase fuel-efficient vehicles and change their fleet, car manufacturers will produce more energy-efficient cars, making these types of vehicles more available to everyone in that market. When multinational corporations decide to eliminate harmful substances like PVC plastics in their products, they greatly decrease the use of this substance by changing contracting specifications with their suppliers.

As in any marketplace, the amount of money a buyer has is the financial equivalent of power, however, even individuals with relatively little individual economic power command attention from manufacturers when they operate as an aggregate greater than their individual money and power.

The inherent contradiction between green consuming and sustainability turns on the fact that sustainability is in many ways a plea to current generations of privileged people to stop consuming so much material, energy, and water, and wasting so many resources in redundant and obsolescing consumer goods. Some have criticized appeals to consumers to consume more "green" goods as trying to buy ourselves into sustainability without the harder work of redefining happiness, prosperity, and security in nonfinancial terms. ***See also* Volume 2, Chapter 4: Green Consumerism.**

References

Dauvergne, Peter. 2008. *The Shadows of Consumption: Consequences for the Global Environment.* Cambridge, MA: MIT Press.

Princen, Thomas, Michael Maniates, and Ken Conca, eds. 2002. *Confronting Consumption.* Cambridge, MA: MIT Press.

Voigt, Christina. 2008. *Is the Clean Development Mechanism Sustainable? Some Critical Aspects.* Vol. 8. Sustainable Development Law & Policy, pp. 15–21.

WASTE, POLLUTION, AND TOXIC SUBSTANCES

Waste is a by-product of human habitation and industrialization. As both have dramatically increased in the last century, so too have the environmental impacts on land, air, and water. Many of these impacts occurred, and are occurring, without knowledge. As knowledge about the environmentally degrading environmental and human impacts develops, laws are passed describing what "pollution" is.

What Is Pollution?

Pollution is a term of art describing illegal and regulated environmental impacts. In the United States, it is narrowly defined by regulations to generally mean emissions of a certain chemical over an allowed amount.

This differs from the general understanding of pollution to be any negative environmental impact. Many industries are permitted to emit a certain amount of a given chemical. (For more information, see www.scorecard.org, the U.S. Toxics Release Inventory by zip code). The limited legal definition of pollution is a controversial issue with implications for sustainable development. Many environmentalists contend that the amount allowed to be emitted is too high and that the list of chemicals regulated is incomplete. The U.S. environmental enforcement system is largely reliant on self-reported information from industry. Of the more than 80,000 chemicals in commerce in the United States, only about 2 percent are tested thoroughly for public health impacts, and very few for ecosystem impacts.

As knowledge about the locations of wastes, the environmental impacts of waste, and the public health impacts of wastes increase public policy engaging these important issues of sustainability will increase. In chemicals that bioaccumulate in humans and that persist in the environment, scientific knowledge and cumulative effects have pushed the policy envelope.

Pollution Prevention

Some technologies clean up waste and store pollution after it has been created. Those strategies are often call end-of-pipe or end-of-smokestack approaches to pollution. These technologies help businesses operate within the pollution limits set by their permits. They do not prevent pollution from being created; they do deal with it afterwards. Cleanup strategies include reuse and recycling, containment and treatment, and even massive cleanups of contaminated lands. All cleanup strategies themselves use energy and other resources like clean fresh water. In efforts to encourage cleanup behaviors like recycling, these calculations are sometimes overlooked.

As a rule, it is true that prevention is far less costly than fixing a problem that has been allowed to occur. That is the wisdom behind the old saying, "An ounce of prevention is worth a pound of cure." Strategies that intentionally minimize or eliminate waste and pollution in the design of a product or process are pollution-preventing strategies. Strategies and technologies that actually design pollution and waste out of the production process and products achieve pollution prevention. Pollution prevention saves not only resources and energy used to make products but the energy and resources used to recycle and reuse them.

Sector-Based Approaches

In this approach to environmental regulation of industry, industry is divided into sectors by the Standard Industrial Classification (SIC) code. The United States and many other developed nations use sector approaches to refine the fit between the environmental permit and the industries emissions. Early environmental regulatory approaches relied on a "one size fits all" approach.

Sector-based approaches seek to help industry comply with environmental laws. They theoretically help government understand the complexities of a type of industry so they can work together to reduce environmental degradation. Some international treaties incorporate the SIC codes in establishing comparable environmental regulatory programs and policies.

From an environmental and community perspective, sector-based approaches are an improvement in most developed countries because they increase regulatory flexibility and transactional transparency. In some industrial nations like the United States, sector-based approaches can reveal if the reach of environmental regulatory authority is enough to act as a platform for new sustainability policies.

References

Hofrichter, Richard, ed. 2000. *Reclaiming the Environmental Debate: The Politics of Health in a Toxic Culture.* Cambridge, MA: MIT Press.

Seliger, Gunther. 2007. *Sustainability in Manufacturing: Recovery of Resources in Product and Material Cycles.* New York: Springer.

Risk Assessment

Risk assessment is a form of analysis of the probability and magnitude of harm from various events and activities. It is widely used to make decisions. Insurance relies on this computation of risk of harm in making decisions about whether to insure, and if so, how much to charge for insurance. Risk assessment is also used in decisions about development projects, and it is used by governments in budgets and planning activities.

Related to the science of risk assessment, risk management determines how to plan for and communicate about risks. Risk perception is a science devoted to examining the qualitative aspects of risk, not simply its quantitative aspects.

Government often requires a risk assessment to be performed in many areas of environmental and developmental activity. These studies are used to set priorities and determine funding priorities. Risk assessment is often done as a matter of expert assessment. However, risk perception is a developing aspect of this science, and a commonly held perception of environmental and ecological risk can provide an important common platform for environmental action that is missing in current controversies. Government has an important stake in developing such common ground as a basis for legislation and regulation for sustainability, especially in changing assumptions of permissions to pollute toward assumptions of eliminating all forms of waste and pollution. ***See also* Volume 2, Chapter 4: Risk Assessment.**

References

Beer, Tom, and Alike Ismail-Zadeh. 2002. *Risk Science and Sustainability: Science for Reduction of Risk and Sustainable Development of Society.* New York: Springer.

Cutter, Susan L. 1993. *Living with Risk: The Geography of Technological Hazards.* London: New York: E. Arnold.

Hlavinek, Petr et al. 2009. *Risk Management of Water Supply and Sanitation Systems.* New York: Springer.

Morello-Frosch, Rachel, Manuel Pastor, and James Saad. "EJ and Southern California's Riskscape: The Distribution of Air Toxics Exposures and Health Risks among Diverse Communities." *Urban Affairs Review* 36 (2001):551.

Robson, Mark G., and William E. Toscano. 2007. *Risk Assessment for Environmental Health.* Hoboken, NJ: Jossey-Bass.

WATER SYSTEMS

Water is the basis for life on Earth. Fresh water is especially important for life of many animals and plants. With human population increases, increases in water consumption, and increases in waste production, the supply of fresh water is an ecological crisis of sustainability. When a country has less than 1,000 cubic meters per person per year of water, the health and economic development of that country are threatened. If the water is 500 cubic meters per person per year, then the ability of humans to survive is in question. In the late 1990s, about 28 nations had water scarcity issues; by 2025, this number is predicted to increase to about 56 countries. The number of people and ecosystems affected will be much greater because of increases in human population and in water use.

Increases in water use are caused by increases in deforestation, agriculture, mining, and human settlement development. The ability of the hydrological cycle to recharge itself and hold and create fresh water is lowered by the loss of forests, wetlands, and estuaries. Climate changes resulting from global warming may also affect the supply of fresh water by increasing ocean levels and increasing floods and hurricanes. ***See also* Volume 2, Chapter 4: Water.**

Water Cycles

Water moves through the air and land in cycles depending on the climate and landscape. Water interacts with the land and the air driven by thermal conditions. The pattern of these interactions can be observed as a repetitive cycle of interactions. Global, regional, and local water cycles occur continuously. Processes of precipitation, infiltration, runoff from impermeable surfaces, runoff generally, evaporation, and transpiration all contribute to the water cycle. Fresh water is reliant on ecosystems to filter it through physical, chemical, and biological filtration processes.

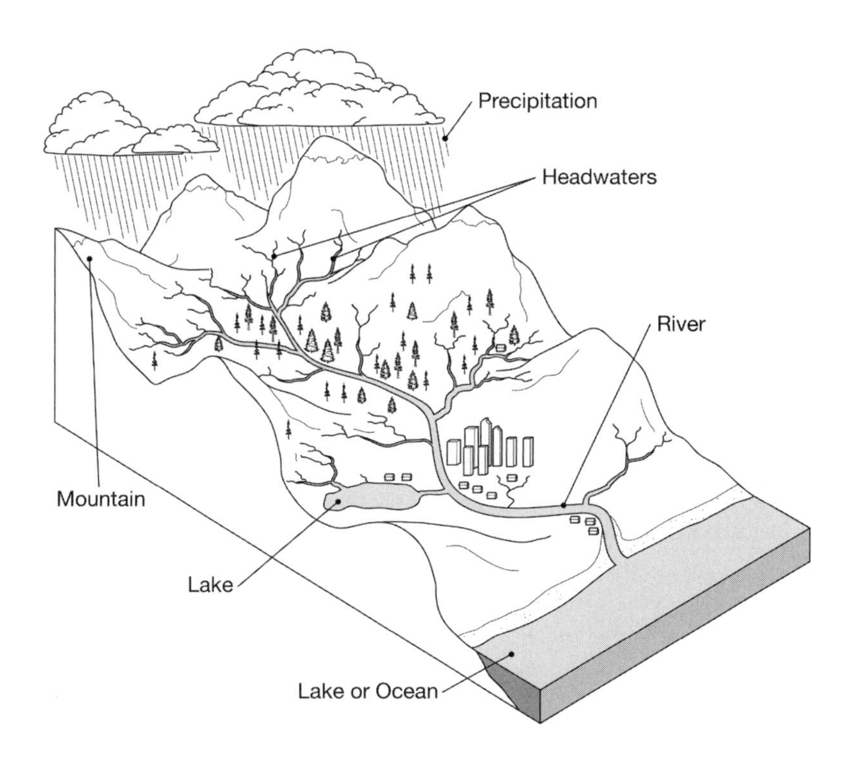

Figure 2.13 • Water collects in a basin by draining downhill from various sources. The land area from which water collects is sometimes called a watershed. Outside of North America, watersheds are used to describe the dividing lines between these basins. A water cycle is the interaction of land, air, and water in a watershed. Illustrator: Jeff Dixon.

Water Shed

Water flows from its sources to its endpoints from a variety of tributaries. Along such flows, water interacts with land and air in cycles. The distinct geographic area of such interactions from source to end is a single watershed. Watersheds are part of emerging measures or indicators of sustainability. Agenda 21 lists adoption of an integrated, watershed-focused approach to water quality as a goal.

Some environmental groups balk at watershed programs because they see problems of effective environmental enforcement. In many nations, and in the United States, the main focus of environmental regulation for water quality was on point sources of chemical discharges. Nonpoint sources, unregulated industries and cities, and development in general, however, contribute large amounts of pollution to the water, causing environmental degradation of natural systems.

Watersheds are part of many controversial issues of water use. As water quality and quantity become one of the first natural systems to erode, they become an environmental indicator. With climate change and the potential desertification of the equatorial tropics, watersheds will receive renewed attention. The use of watersheds is thought to increase accountability for actual water usage and water practices. The U.S. EPA allows people to surf their watershed.

In many places, agricultural industries consume the bulk of available water. In many urban areas in developing nations, water conservation is seldom practiced despite decreases in quality and quantity.

In 1996, the U.S. EPA broke down threats to aquatic species by watershed. They found at least one aquatic species to be at risk in 403 watersheds, between two and five species at risk in 745 watersheds, and more than five species at risk in 422 U.S. watersheds.

Runoff

Water flowing from land toward water is runoff water. This water carries with it many kinds of material. Dissolved material within the flow of this water is also called runoff. One of the contexts for runoff in sustainability relates to the intensification of agricultural usage of fertilizers and pesticides. These run off the land into rivers and eventually into the sea. For example, the death of the coral in the Great Barrier Reef of the east coast of Australia is due in part to excessive nitrogen and phosphates that is found in currently used agricultural products and processes. This type of runoff can cause "dead zones," or areas around the deltas of polluted rivers where no life exists. The number and size of dead zones have been increasing, and they are an area of study by environmentally based sustainability advocates.

Runoff also results from paving over land. The hard human landscapes of roads, streets, parking lots, airports, and malls are usually paved with hard surfaces. The water that would normally percolate into the ground and be slowly released into the ecosystem instead runs off the hard surface to wherever it is channeled. It is often channeled into public formal or informal sewer systems.

Some sustainable communities are exploring alternatives to nonporous surfaces. Other communities are exploring the idea of green roofs in the context of water conservation. Rainwater conservation is not a new idea itself. Many arid locations try to capture rain in various containment systems, both above and below ground. A green roof uses plants that grow on the roof to contain water. It requires that the roof be waterproofed and able to repel roots. The green roof also needs to have a good drainage system. They are popular in Europe but just starting in some places in the United States. They can be modular or interlocking. The building structure of the roof itself must be strong enough to support the additional weight. In terms of water retention, green roofs can hold storm water and drain it more slowly. This helps prevent storm water sewer overflows. Many U.S. cities have consolidated sewer systems that overflow the waste treatment plants when storm water flows overtake their capacity. A green roof moderates the flow of water from rainstorms so that this does not occur and thus preserves fresh water. Storm water overflows from older urban consolidated sewer systems are a source of groundwater contamination. A green roof can also transfer water to the immediate environment through processes of evaporation and transpiration. This can have a cooling and cleaning effect on the air around the building. In most summer locations in the United States, green roofs can retain 70 to 90 percent of rain and between 25

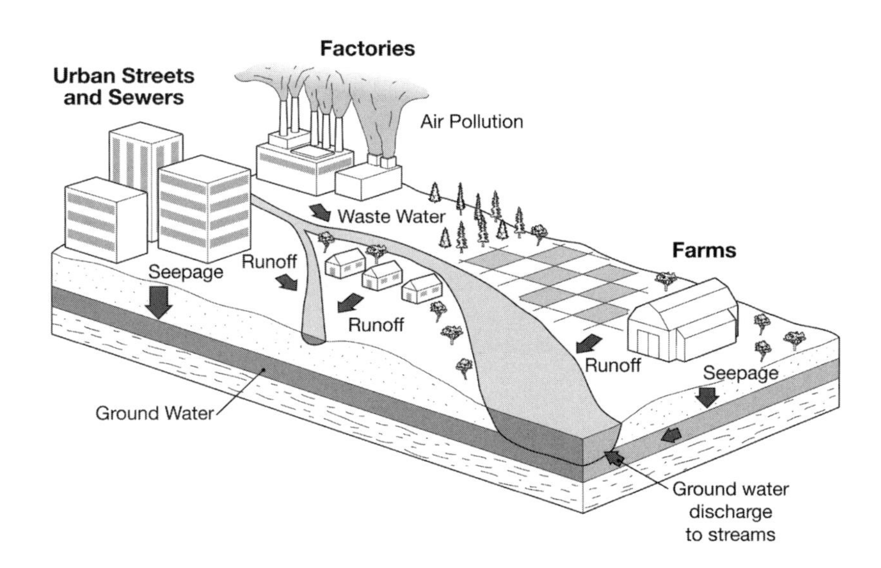

Figure 2.14 • Groundwater contamination can come from sources located on land such as factories, farms, and urban streets and sewers. Air pollution can also contribute to groundwater contamination as it falls to Earth dissolved in precipitation. Illustrator: Jeff Dixon.

and –40 percent in the winter. The amount is dependent on the types of plants that are grown.

The city of Portland, Oregon installed its first green roof in 1998. In 2008, about eight acres of roofs in Portland were green. The city's goal is to increase this amount to 51 acres out of the 12,400 acres of rooftops in Portland. The city calculates that this prevents about 50 percent of storm water runoff. The city offers an incentive program called Grey to Green Grants to help offset the higher cost of installation of green roofs. It is estimated that green roofs last twice as long as shingle roofs in Portland's rainy climate.

Groundwater

Lakes, rivers, and streams may exist wholly or partly underground. These water sources are called groundwater. Water percolates or seeps down through the soil into underground aquifers and streams. This recharges the aquifer with fresh water.

The rate of groundwater recharge is of keen importance to sustainability planners. Water is a basic system on which life depends, and it needs to be sustainable. Parts of much new ecosystem analysis include the recharge rate of the groundwater. If the recharge rate is too low, the water can become tainted with toxic pollutants. A concern of many environmentalists is that the recharge rate has been too low for too many years and that the environmentally degrading land use and industrial process of the past are seeping into the groundwater.

Hydrogeologists are currently mapping the course of groundwater. Many of the largest aquifers are known. The next step is to determine the present and future water quality of the groundwater.

CHAPTER 3

Government and United Nations Involvement

ROLE OF GOVERNMENT IN MOVING THE ECONOMY TOWARD SUSTAINABILITY

Government has a role to play in organizing human activities. The challenge of sustainability to governments is to define its role in achieving environmental and ecological sustainability. What actions and values can a government foster to ensure that human economic activities and human communities live respectfully within the limits of the webs of life supporting all life on this planet?

Regulation, Taxes, and Subsidies for Transition to Clean Production

Green Taxes, Carbon Taxes, and Green Fees

Taxes function by adding costs to an activity or product. Adding such costs makes them more expensive and less desirable. Rational tax policy generally adds such costs to things we do not want to encourage. Subsidies are usually added to things we do want to encourage. Tax policy for sustainability would add costs to the unsustainable aspects of goods and production methods and subsidize the sustainable aspects. This would encourage businesses to reduce or eliminate the unsustainable aspects of their operation.

Taxes on hydrocarbon produced by businesses could be used to discourage dependence on fossil fuels. This, together with subsidies or tax credits for renewable fuels, could shift our economy away from its use of unsustainable energy resources.

Green Fees and BTU Taxes

Fees function in the same way as taxes to raise the costs of production. Taxes require legislative action before they can be imposed on taxpayers. Unlike taxes, fees can be imposed by agencies and other bureaucrats without legislative approval. For this reason, fees are less politically controversial to a broad segment of the population, but they can be very controversial to the specific businesses that they affect. Both fees and taxes may be passed along to the ultimate consumer of products and services in the form of higher prices.

Fees are often used to pay for the cost of a particular service. Green fees are used to pay for costs associated with recycling of waste. These fees can also be used to raise capital for changes to infrastructure and buildings, maintenance, and other fixed costs that allow organizations to adopt more ecologically responsible products and practices that cost more than environmentally damaging low-cost competitors. In the latter sense, fees operate as a subsidy to encourage positive changes in behavior. When these fees are imposed on sustainably useful and desirable behaviors, however, they may be counterproductive. ***See also* Volume 2, Chapter 2: Market-Based Strategies; Volume 2, Chapter 4: Market-Based Solutions to Environmental Degradation.**

Setting Standards That Force Technology

Legislation and regulation often set standards for economic and social activity. In terms of environmental protection, current law has mostly set standards in terms of allowing some forms of pollution and waste as a cost of economic and social activity. Setting the best amount of waste/pollution to economic benefit (sometimes called an optimal amount) has fostered a sense of the right to pollute in the form of permit or permission to pollute instead of setting standards (and goals) aimed at elimination of waste and pollution.

Government has a powerful role to play in setting standards that challenge human activities to prevent and eliminate waste and pollution. The effect of setting standards for permissible pollution at low levels or zero forces companies to alter their technology to achieve that goal where such technology is available. Where technology is not currently available, setting such standards to become effective by a specified future date creates an incentive to develop new technologies.

Fuel Economy Standards

Governments may set goals of fuel economy that are enforced by standard setting and also through measurements of fuel economy. The U.S. federal government has not joined with other nations in setting fuel economy goals that would reduce greenhouse gas emissions and global climate change, but some state governments have tried to do so only to find their efforts overridden by the federal government under the Bush administration. In one of its earliest acts, the Obama administration allowed states like California to adopt stricter fuel economy standards than the national standards for automobiles.

Reference

Knechtel, John, ed. 2008. *Fuel.* Cambridge, MA: MIT Press.

Recycling of Building Materials

Governments can force changes in manufacturing processes and products by mandating that materials be recycled rather than sent to a landfill. The majority of nonhazardous solid waste that enters a landfill is demolition and other solid waste from construction sites. These include building materials that have been taken out of existing built structures such as concrete, metal, wood, and other materials. Much of this material is capable of being reused provided there is some level of processing. The cost of material processing may be prohibitively expensive and inefficient unless a large enough volume of material is handled, allowing the processor to attain economies of scale in its operations. When governments mandate recycling of building materials in a high proportion, this drives a market for recycling existing materials. It also encourages the production of new materials that are cost effective to reprocess.

Some local governments are now requiring substantial reuse and recycling of building materials to keep their landfill sites from filling up as quickly and avoiding the cost of early replacement. Local governments, like counties, have the duty to pay for the cost of garbage services, and their taxpayers will have to support these costs through their taxes. By requiring increased recycling of building materials, some of these costs are reduced and shifted onto those who can avoid these costs by increased use of recycling.

Regulating Runoff on Urban Lands: Minnesota Legislation

A combination of standard setting and emissions control is to regulate runoff from the built human environment. Local governments exercise a measure of control over the location of businesses locally through their land use planning and zoning rules, and regulations and policies. These laws can have a dramatic effect on the appearance of the built environment and its interface with the natural environment.

Runoff from the built environment contributes many sources of pollution to surrounding waterways, including underground water sources. Pollution can include sewage, fertilizer, pesticides, and even prescription medications. These can contribute to the serious degradation in local water quality. Minnesota has adopted an urban policy of requiring all new structures to contain and filter runoff water from buildings before it is released into sewers or other public infrastructures. This has had the effect of encouraging the creation of "green roofs" or roof gardens and other greening of streets, sidewalks, driveways, and roadways to filter runoff water.

Government as Consumer: Contracting

Governments exercise unique power to influence voluntary private marketplace transactions without the need for legislation or regulation. This power is often exercised through government acting in its role as a purchaser of products. The specification of sustainable products and materials in government procurement can create a market for technology and materials that stimulates investment and innovation. Without such a stimulus, the costs of innovation might deter changes in technology and materials in many sectors of the economy.

Department of Defense Contracting and Sustainability

Government as a consumer can be a powerful marketplace incentive to achieve widespread changes in business behavior. The Department of Defense is one of the largest spenders in the U.S. economy. It is also the largest consumer of fossil fuels and one of the largest construction enterprises directly and indirectly through contracting with private businesses. The Department of Defense has initiated several policies to implement a commitment to sustainability, including a requirement that new construction satisfy LEED (Leadership in Energy and Environmental Design) certification standards.

This standard will have an impact throughout the global economy as manufacturers of products and service providers compete to meet it. The effect of such competition will increase the availability of compliant products throughout the environment at prices much lower than initially would be expected. In addition, when such standards become the expected standard across an industry or business, it establishes expectations for all members of the industry or business. Because these effects are the result of individual choices to compete in the marketplace, they do not require enforcement or supervision. In addition, voluntary choices are much quicker to implement than are involuntary choices. Enforcement and compliance mechanisms, like investigators and police, and lawyers are expensive and time consuming. They are costs of governmental programs that rely on force to establish compliance with standard setting.

Government can sponsor educational programs on a broad spectrum from preschool to adult education. Education is an essential part of any program of change. The change of business and economics toward sustainability requires redefining some basic approaches to profit, private property, ethics, and widespread understanding of our fragile, valuable ecosystems on which all life depends. Governmental programs can help change our fundamental approaches to these matters in ways that create champions and constituencies for sustainability within business and industry.

Reference

Durant, Robert F. 2007. *The Greening of the US Military*. Washington, DC: Georgetown University Press.

Ray C. Anderson: Leadership for Change toward Sustainability

Ray C. Anderson was born in West Point Georgia. He became an industrial engineer by training before founding Interface Corporation in 1973. Interface manufactures carpeting. The business operated in a traditional way, using petroleum-based products and creating a large environmental footprint with products leading to landfills until 1994. That year Mr. Anderson, chair of the company he founded, read *The Ecology of Commerce* by Paul Hawken. Anderson says that it changed his life and his company's life. The dramatic story of his conversion of his company to sustainable business practices is told in the film, *The Company*. He has become one of the most ardent, green corporate leaders working publicly and tirelessly to convert other businesses and organizations to sustainable business practices.

References

Ray Anderson. www.interfaceinc.com/goals/sustainability_overview.html.

Bakan, Joel. 2004. *The Corporation: The Pathological Pursuit of Profit and Power.* New York: Free Press.

Facilitating Multistakeholder Dialogue and Partnerships

Government also has a unique power to call people together to discuss common challenges. This is the power to convene interested parties, sometimes called stakeholders, for a purpose that affects all, but is rarely discussed with all interested parties present and participating. This approach to making public policy sustainably is discussed extensively in Volume 3.

References

Hemmati, Minu et al. 2002. *Multi-Stakeholder Processes for Governance and Sustainability.* London, UK: Earthscan.

Sabatier, Paul et al., eds. 2005. *Swimming Upstream: Collaborative Approaches to Watershed Management.* Cambridge, MA: MIT Press.

The Importance of the Bureaucracy: Administrative Competencies

Governmental policies are implemented through agencies, administrators, and bureaucracies. In a democracy, agency powers are broadly described in legislation that delegates to the agency the power to pursue certain limited goals. In other forms of government, administrative agencies have the same powers, but the source of their authority would be different, for example, a monarch.

The success of a particular policy often depends on the way that a bureaucracy chooses to view its powers and pursue its goals. Administrators are important in shaping the views and goals for their agencies. The skills and competencies of agency administrators become critical in shaping how they undertake fulfilling the mission of their agency. In

The Common Sense Initiative

From 1995 to 1999, the EPA hosted an experiment in multistakeholder dialogue called the Common Sense Initiative. The EPA invited key stakeholders to participate in a series of dialogues to find alternatives to current governmental regulations that were "cheaper, cleaner, and smarter." Stakeholders included labor, environmental advocates, environmental justice groups, state and local governments, as well as the EPA. These dialogues brought together representatives of the three elements of sustainability in a unique process, allowing for direct communications between people and organizations that rarely communicated in a constructive, face-to-face dialogue. All decisions were to be reached by consensus, although each sector committee could define the meaning of consensus for its processes.

Certain industrial sectors were selected to participate in this experiment based on their permitted emissions and other factors. Some of these sectors were primarily small businesses; others were dominated by large, multinational corporations. This sector-driven approach is a very different approach to environmental protection from the traditional media-based approaches of the major statutes that the EPA enforces such as air, water, and solid or hazardous waste deposited on land. Six industrial sectors were selected to participate in the dialogue: automobile manufacturing, computers and electronics, iron and steel, metal finishing, petroleum refining, and printing.

The six sector committees produced 43 different projects and led to other initiatives within the EPA including the Sector-Based Environmental Protection Action Plan and the Stakeholder Involvement Action Plan. Some projects focused on issues of pollution prevention and product lifecycle management. Others tackled issues related to monitoring of fugitive emissions and other information management challenges. One of the most comprehensive projects was designed by the printing sector. Its project, called PrintStep (Simplified Total Environmental Partnership), restructured permitting processes to promote pollution prevention and community involvement.

shaping policies for sustainability, government bureaucracies will find that they are expected to adopt leadership positions and demonstrate competencies in constructing enforcement, compliance, and public policies that balancing goals of sustainability that may at times compete and conflict with each other.

BARRIERS TO GOVERNMENTAL ROLES IN SUSTAINABILITY

Despite all the powers government has to move toward sustainability, a few key factors can make these goals unachievable. These barriers are different for different countries and at different times. A few types of barriers are recognizable in many situations. In developed countries, a move toward sustainability will cause some industrial sectors to lose money, impose some costs on individual businesses and consumers, and require leadership that motivates these parties appealing to shared, long-term goals and a sense of urgency. In less developed nations, moving toward sustainable development will confront problems of political corruption.

Governments do not control all environmental actions. In many nations, the role of the government is limited because of the lack of resources available. Many of these governments seek to increase the quality of life of their citizens no matter what the cost to the environment. In other nations the role of government is limited because forces outside the government are stronger. Some nations are in a state of civil unrest or war. They too are often without resources. Without adequate resources to study the environment and know ecological baselines, many governments cannot achieve sustainable development and could irreparably damage their environment. Some governments may have political values that limit government intervention to some, most, or all matters. In many instances, however, the so-called private sector is actually supported by public laws, government enforcement of market rules, and subsidies.

Governments with resources may view the environment as a political issue and therefore subject to political negotiation and compromise. Some reject the notion of sustainability as a liberal issue that contradicts their understanding of the environment as a source of unlimited growth. Some governments are hampered by the limitations of their use and application of science. Governments that politically compromise the environment for their short-term economic gain also do not study the environment enough to know their ecological baselines and carrying capacities and may risk irreparable damage to their environments for future generations.

All governments have value structures that limit their approach to the environment. Private property values as a basis of strongly held liberty values may impinge on governments that seek to know the ecological carrying capacity in areas that transcend the boundaries of private property ownership. Boundaries of private and public property, along with borders of nations, blur the analysis of natural systems. Natural systems respond to other aspects of their ecosystem, not human-imposed boundaries.

Borders and Jurisdiction

The government of a country can exercise power over its territorial areas defined by its physical borders. These borders may not have a relationship to ecosystems or other environmental features. They may reflect other facts of history without regard for the geographical or cultural features of the place. Ecosystems and their features, like watersheds or mountain ranges, often cross borders between countries.

Some continents, such as Africa, North America, and South America, were colonized by other nations. Many colonizing nations sought natural resources such as salt, gold, and oil and divided the land into parcels to efficiently retrieve these resources. In the process many indigenous people were killed and relocated, sometimes intentionally and sometimes unintentionally, when diseases were introduced for which they had no resistance. The land was divided by borders that were defended and guarded. Some areas with known natural resources and viable transit routes were heavily exploited. Other areas with few natural resources and nonviable transit routes were often left alone. This global process of colonialization has left ecosystem impacts that may affect modern efforts of sustainability.

In a modern political state, pollution can occur within one state or nation and be carried into another state or nation. Airborne pollution can travel long distances, as in the case of acid rain. When pollution travels it can enter other parts of the ecosystem, such as leaving the air to fall as rain onto the ground, and from there it can move into plants and animals. Environmental protection systems based on nations or states can facilitate greater environmental degradation simply by moving regulated pollution into unregulated areas. Nations or states seeking to attract economic growth through industrialization may do so by environmental deregulation or lack of real enforcement of environmental rules. This is a process known as "race to the bottom."

Border and property boundaries that do not reflect ecological systems in nations that have little or incomplete environmental regulation pose a significant challenge to sustainable development. A foundational component of the emphasis on land division and human ownership of land is private property.

References

Conca, Ken. 2005. *Governing Water: Contentious Transnational Politics and Global Institution Building.* Cambridge, MA: MIT Press.

Yang, Zin. 2008. *Strategic Bargaining and Cooperation in Greenhouse Gas Mitigations: An Integrated Modeling Approach.* Cambridge, MA: MIT Press.

Ownership of the Environment: Private Property

Ownership of property in some cultures gives individual owners the power to determine what choices to make regarding that property. This right may exist without knowledge of, or regard for, the consequences of that individual choice for the ecology of the place. To the extent that government ensures such absolute expectations of rights to property, sustainable choices for environment and community may be frustrated.

In the United States, 3 percent of the land is owned by industries and 2 percent is owned by private residences. Overall, in the United States there are about 2.1 billion acres, which is about 7.7 acres per person, 14 times higher than the rest of the world and three times higher than any other industrialized nation. About 16 percent of that land is occupied by 341 Metropolitan Statistical Areas. The federal government is a large landowner, owning 96 percent of Alaska, 86 percent of Nevada, 66 percent of Utah, 63 percent of Idaho, and 52 percent of Oregon, as well as substantial acreage in other states. This represents about 730 million acres, or about one-third of the total land and water area of the United States. The ability of the state to control the use of land, whether public or private, leased, owned, or held in trust, is important for any policy of sustainability. ***See also*** **Volume 2, Chapter 4: Private Property.**

Reference

Emerton, Lucy et al. 2006. *Sustainable Financing of Protected Areas: A Global Review of Challenges and Options.* Washington, DC: International Union for Conservation of Nature.

Real Estate

Information is essential in making sustainable choices. Government has the power to assemble information on a wide array of topics in order to perform its duties. Information about the environment is assembled in many offices of government. Sometimes that information is publicly available. Timely public access to environmental information is essential knowledge for residents. Limits are regularly placed on the ability of the public to access information gathered by the government. One significant constraint on public access to governmental information gathered about the environment is the ability of businesses that discharge waste and pollution to shield their activities from public scrutiny to the extent that it involves confidential business information. This is information that the business asserts must remain secret in order to protect their trade secrets. Another significant constraint on public access to governmental information is national security. This involves information that the government asserts must remain secret in order to protect the nation's own health, safety, and welfare. These assertions of a right to secrets can prevent the true and full accounting for information about what is in our land, air, and water. ***See also*** **Volume 2, Chapter 4: Real Estate.**

Business and Trade Secrets

Sustainable development attracts many entrepreneurs, individuals, and businesses seeking new ways of doing business in a sustainable manner that still makes a profit. In a sense, sustainable development is a new growth industry. As such new processes and products are created. Traditionally, these are protected by the legal process of patent and trade secret protection. Sustainable development is under particular scrutiny because if a new process is sustainable, traditional business methods of protecting it may contravene public policies of disclosure necessary to protect the environment and the threat of irreparable damage that would impair the ability of future generations to enjoy it.

Patents and trade secret protection are meant to encourage private businesses to do the research and development necessary to move society forward. By giving special protection to the first business to market with a new good or service, public policy protects initial profits to cover the costs of research and development. Patents of life are particularly controversial because they could have unintended impacts on the environment in ways that jeopardize future sustainability. Patents of traditional knowledge of indigenous peoples are another highly controversial area.

Patents and trade secrets are not self-operating. They require the business person to take affirmative steps to claim the product or process as their own. Trade secrets require that the information have commercial value, and that the information or how the information is used be a secret. The potential owner of the trade secret must take reasonable steps to ensure its secrecy. This can mean restricting access to its research facilities, preventing all publication and dissemination of information related to the secret, and requiring employees who work on the secret to sign nondisclosure agreements. In the case of sustainability secrets, much of this information can be environmental. It can relate to new crops, new ways of remediating environmental areas of toxicity, new ways to construct buildings and roads, or new ways to create energy. This is a problem in sustainable development because the idea seeking trade secret protection may need to be disclosed to be useful. It simply may not be able to receive trade secret protection. This is one argument that many sustainability advocates use to promote greater governmental intervention.

The other traditional method of idea protection is the patent system. Nations differ significantly in their approaches and levels of enforcement of patents. Under patent protection, the entrepreneur must disclose the idea, and only if that idea is new and novel does the government patent office issue the patent. Generally, an idea that is not new and is used for one year loses the ability to get a patent. If a patent is pending while the U.S. Patent Office is considering the patent application, more time is granted.

Sustainable development as a public policy and private profit-making idea is problematic. This problem was the creative push for the Designers Accord. Designers face all types of environmental challenges. Not only do designers design buildings and follow LEEDs design criteria to be sustainable, they design cloths, products, manufacturing processes, and art. They use plants, inks, and chemicals in processes that create waste and pollution. The Designers Accord is a set of voluntary business principles that tries to make sustainability an early part of the design process. It has been signed by more than 3,500 designers, some of the biggest design firms, and endorsed by the major design trade associations. These accords were created by IDEO, a business design firm founded in 1991. IDEO is a design and consulting business that uses a human-centered and design-based approach to help organizations innovate and grow. Sustainability is made an early part of the process by speaking with every new client about it first, and by sharing knowledge about sustainability with others in the industry. In this way it challenges the traditional notion of trade secrets and patents and instead shares best practices based on shared experiences. Reluctant clients and lack of environmental literacy by designers have been impediments to sustainable development in the design industry. The Design Accords seek to alter this problem. The guidelines to implement the accords encourage proactive dialogues with all clients about the environmental impacts of their work and sustainable alternatives. Designers are to share knowledge with others in the industry to gain a competitive advantage, instead of the other way around. ***See also*** **Volume 2, Chapter 4: Technology Transfer, Intellectual Property Rights, and the Developing World.**

Military Secrets

Military groups all over the world have significant environmental impacts. It is part of the nature of war to create, or threaten to create, environmental impacts that may irreparably affect systems of nature on which future life depends. Military organizations use many types of materials and processes aside from war that are used in the defense of nations. If these materials and processes are known to outside groups, then national defense is weaker. The need to keep them secret is based on national defense, even though they may have significant environmental impacts.

Regular military operations have environmental impacts. The routine deployment of troops in training and mobilization exercises involves the use of petrochemicals. Weapons testing creates severe environmental impacts and may leave heavy metals in the land and water. Waste disposal is a large issue and often must be secret. Building, closing, and realigning military bases can have environmental impacts. For example, when an airfield on a base closes, the toluene used for wing deicer often seeps into the water systems of nearby communities.

The military is becoming more open to sustainability. In November 2008, the U.S. Army released its first annual sustainability report. The report covers the period from 2004–2007. A total of 16 army installations have comprehensive Installation Sustainability Plans. Also, 78 percent (301) of the Army Construction Projects in 2007 were designed to LEED new construction sustainability standards. All of the army's 161 installations have an environmental management system, and 31 percent of them apply the International Standards Organization standards. They have decreased energy use by 8.4 percent. The U.S. Army still generates large amounts of waste. From 2003 to 2006, there was a 35 percent increase in hazardous waste generation and an 8 percent increase in pounds of hazardous waste generated per $1,000 net cost of army operations. Overall, there was an 11 percent increase in toxic release inventory during this time frame.

The concern for sustainability pushed the army to report and plan around sustainable development principles, even in the face of a tradition of military secrets. The express purpose of the army's first report is to inform and engage communities, business, and other interested stakeholders on their status and progress in advancing the principles of sustainability in operations, installations, systems, and community engagements. The army expressly embraces the Triple Bottom Line of Sustainability by aligning its missions to environmental stewardship, community health, and economic growth. Their first report will serve as a baseline for future measures of sustainability. (For the full report, see www.aepi.army.mil.)

Military organizations will still face environmental controversies and the need for secrets. Over time, the environmental impacts of any secrets become known because the impacts begin to accumulate in the environment. The effects of nuclear testing in the Bikini Islands became known as attempts made decades later to make them habitable uncovered a much larger extent of ecosystem damage. The effects of weapons testing became apparent in Hawaii and Puerto Rico as the environmental impacts became known. Military technology also creates environmental impacts with unintended impacts that may threaten environments. Military technology is often one of the most closely guarded secrets. From laser-guided nuclear missiles to coal-fired jet bombers run by computers, the potential threat to sustainability becomes greater as these weapons become more powerful. Research and development, and sometimes distribution and creation, remain secretive. Older military technology does not just go away. Although it may no longer be a secret, it becomes used in other countries. There it may also have powerful environmental impacts that can threaten sustainability.

Inertia: Sunk Costs, Old Technology, Political Will

Policy decisions are implemented by both action and inaction. Inaction is a choice to keep things as they are, even when the failure to act will predictably lead to a state of systemic chaos. Sometimes this tendency toward chaotic states is called entropy. Many factors in human economies can paralyze action, even when inaction probably will result in collapse or chaos. Factors contributing to such inertia include uncertainty, sunk costs in existing infrastructure now paid for and completed, and the political costs of change.

Governmental Corruption

Political corruption is present in many governments and in all types of other organizations. In many developed countries, we take for granted that people who work in the public sector do not expect tips or payments for doing their jobs. That assumption is not true for many other countries, where public employees at all levels may be involved in conduct that uses public power for private gain. In most governments there are laws against taking bribes, stealing money meant for public uses, and using governmental force to extract payments from people. These laws are not always enforced equally. When moving toward sustainable

Transparency International

Transparency International is an organization dedicated to monitoring corruption. The organization defines corruption as "the abuse of entrusted power for private gain. It hurts everyone whose life, livelihood or happiness depends on the integrity of people in a position of authority." It publishes the Corruption Perceptions Index, ranking more than 150 countries in terms of abuse of governmental powers. These rankings are based on expert opinions and opinion polls. Governments high on the list of corrupt governments rank low in rankings of sustainability. Transparency International also publishes The Bribe Payers Index and Global Corruption Barometer.

development, however, accurate evaluations of environmental impacts are necessary.

Moving toward sustainable development in developing countries requires substantial investments in roads, bridges, ports, and other infrastructures. It will also require substantial investment in technology and intellectual property. All of these investments require transfer of substantial sums of money. If that money does not reach its intended uses, development will fail.

Unregulated Market Capitalism

Markets have been very successful at creating wealth, although the distribution of that wealth is inequitable. Markets have arguably been disastrous for ecosystems. The demand to buy, sell, and trade goods for profit will not necessarily respond to other values like the elimination of poverty or protection of the environment on its own. Values like equity and environmentalism are to a great extent intangible and not incorporated in contemporary costs or profits. To the extent that profits do not reflect the value of being an equitable or environmentally responsible society, markets should not be expected to deliver those results. Today's market prices do not reflect the true and full costs of using resources on our ecosystems or our communities. Many of these costs are paid by others including taxpayers, public health expenditures, and nature's own degrading life systems. When governments protect wasteful market behavior, they promote inefficiency in the free market. The government protection of polluters, which is waste, takes away its "free" market status. Government protection of wasteful capitalism and pollution can take the form of subsidies, lack of enforcement of environmental laws, and lack of transparency in environmental transactions.

Rules and regulations that force or encourage sustainable behavior can greatly assist the attainment of sustainable goals, in conjunction with voluntary choices and leadership. Such regulation and voluntary leadership are essential in shaping market results, especially in the absence of full knowledge and accountability for costs.

This absence of knowledge and accountability is another reason for requiring that business and industry comply with the principles of precaution (see Volume 1).

References

Eckersley, Robyn. 2004. *The Green State: Rethinking Democracy and Sovereignty*. Cambridge, MA: MIT Press.

Kraft, Michael E., and Sheldon Kamieniecki. 2007. *Business and Environmental Policy: Corporate Interests in the American Political System*. Cambridge, MA: MIT Press.

Information Constraints

Governmental action, both in terms of setting rules and enforcing them, can be frustrated and ineffective by a lack of knowledge. When government lacks baseline information about pollution and waste from an economic activity, it cannot effectively confine set standards for sustainable, supportable uses. In some cases, we have excluded or exempted businesses from revealing information about their potentially polluting emissions. Sometimes businesses are allowed to keep information secret because it is important to their competitive advantage. Some organizations, including government itself, are allowed to keep information secret if it is related to national security. Many statutes do not require businesses to report pollution if it is under a certain threshold that businesses calculate and self-report.

Information is constrained by national security and confidential business information. Confidential business information is information that businesses are allowed to keep secret because it is important to maintain a competitive advantage. These exclusions can result in drastically incomplete information about what economic activity is doing to its surrounding ecology. This lack of information undermines the accuracy of ecological predictions those economies and businesses rely on about weather, water resources, and climate change effects.

Private Property/Permission to Pollute

Sustainability may require major changes in the way individuals use their private property. Some countries encourage expectations of virtually unlimited rights to use private property according to the wish and even whim of the owner. The expectations are rarely enforceable, even in these countries. The right to engage in activities that damage ecosystems would not seem to be a reasonable expectation; however, under the expectations of absolute rights in private property, some people feel entitled to use their property in ways that harm ecosystems and future generations.

Early environmental protection laws created systems of regulation of private property based on permits. Permits engage private property expectations by permitting a certain amount of pollution. This encourages assumptions of rights to pollute rather than limitations.

Cost of New Technology

One of the most urgent tasks facing businesses and industries is to change their methods of production from dependence on fossil fuels that release greenhouse gases to clean production technologies. These new technologies will cost money as businesses replace fundamentals of their infrastructure. Many businesses simply do not have the money or the ability to borrow sufficient sums of money to change their fundamental infrastructures. In the absence of such capital, whether in developed or developing countries, businesses will not change their business production methods in ways that decrease their environmental impacts.

Capital will have to become available in order to stimulate this type of change. Governments can provide this kind of capital directly through lending programs or indirectly by channeling private investments toward this kind of lending by providing tax benefits for qualifying loans.

Reference

Hess, David J. 2007. *Alternative Pathways in Science and Industry: Activism, Innovation, and the Environment in an Era of Globalization.* Cambridge, MA: MIT Press.

Types of Governments and Economies

The power of government to command changes toward sustainability without the need to achieve popular consensus about measures to be taken would appear to be a substantial advantage in the ability to move quickly toward the goals of sustainability. Democracy as a form of government might be slow to achieve sustainable changes rapidly because of the requirement to submit leadership to a vote at regular, short intervals of time. In modified democratic forms of government, popular mandates or the mandates of legislators can bring down a government whose decisions are unpopular. This means that policies that cause discomfort among well-organized constituencies may be paralyzed when it comes to decisions that cause short-term losses or major shifts in costs, taxes, or subsidies in order to achieve some of the goals of sustainability. This points to the need to groom constituencies and leadership for sustainability in businesses and industrial sectors that will have to make significant changes in methods of production, products, decision making, and accountability in order to achieve the goals of sustainability.

In addition to the form of government, the type of economy that a country has might make significant difference in its ability to make difficult policy changes quickly. In centrally planned economies such as communist or socialist economies, economic planning can occur based on a coherent administrative mandate that affects multiple sectors of an economy without necessarily requiring the input and cooperation of many individual constituencies.

Many indices of sustainability rate countries according to different factors deemed important to achieving sustainability. Considering the form of government and the type of economy of the top performers might provide some insights into how these factors impact the achievement of sustainability.

Yale and Columbia Universities have produced one such index since 1999. The latest environmental performance index for 2008 rated the top 10 nations as follows:

1. Switzerland
2. Norway
3. Sweden
4. Finland
5. Costa Rica
6. Austria
7. New Zealand
8. Latvia
9. Colombia
10. France

Sweden, a constitutional monarchy, regularly appears in the top 10 of sustainability indices. Sweden's King, His Majesty Carl XVI Gustaf, helped to found the Natural Step organization and has promoted leadership in sustainability from its earliest days. Monarchy has played a similar leadership role in Bhutan, where Jigme Singye Wangchuck, formerly king, developed a movement around measuring development in terms of happiness, not dollars. His concept of gross national happiness is significantly different from other measures of development and important to the transformation of economics toward sustainability. Clearly, monarchs can be important. Eight of the top ten sustainable nations, however, were democracies of some type, without monarchs. Democracies included here count republics and mixed forms of parliamentary and directly elected rulers.

When it comes to the type of economy, all of the top performers in terms of sustainability are free market, or mixed free market economies. Mixed economies include those with state ownership of some large businesses or industries. Centrally planned economies, such as China, regularly appear lower on indices of sustainability. ***See also* Volume 2, Chapter 5: Sustainability Indices.**

References

Barry, John, and Robyn Eckersley, eds. 2005. *The State and the Global Ecological Crisis.* Cambridge, MA: MIT Press.

Ho, Mun S., and Chris P. Nielison. 2008. *Clearing the Air: The Health and Economic Damages from Air Pollution in China.* Cambridge, MA: MIT Press.

Josephson, Paul R. 2005. *Resources under Regimes: Technology, Environment, and the State.* Cambridge, MA: Harvard University Press.

Yanful, Earnest K. 2007. *Appropriate Technologies for Environmental Protection in the Developing World.* Boca Raton, FL: Springer Press.

UNITED NATIONS

The idea that nations should come together for any purpose other than trade or war was a radical one for most of human history. Many nations preferred to be isolated and defended their borders from all encroachment. President Franklin D. Roosevelt first used the term "United nations" in the Declaration by United Nations on January 1, 1942. This declaration occurred during World War II when 26 nations agreed to continue fighting against the Axis powers.

Nations had begun to meet to try to negotiate peace before that time. The International Peace Conference was held in The Hague in 1899 to stop wars and develop rules for war. From this conference the Convention for the Pacific Settlement of International Disputes and the Permanent Court of Arbitration were formed. Before the United Nations was the League of Nations. This was established under the Treaty of Versailles in 1919 after World War I. The purpose of the League of Nations was to prevent wars and settle international disputes. When World War II began the League of Nations ended.

It was not until 1945, when 50 countries attending the United Nations Conference on International Organization in San Francisco met to write the United Nations Charter, that the United Nations was formed. The United Nations Charter was developed by some of the most powerful nations in the world on June 26, 1945. After a significant number of nations signed the charter and ratified it, the United Nations was created on October 24, 1945. Today, October 24 is celebrated as United Nations day.

The Role of the United Nations

The United Nations is not a government in the strict sense; it is an association of governments. Together the member states explore ways in which they may act collectively on issues affecting global peace and security. Sometimes, the way the UN functions is to assist member nations in articulating norms of behavior. When these norms are not written into enforceable treaties, this kind of activity is called "soft law." Soft law can be operated on voluntarily by individual groups and by nongovernmental organizations. Hard law is an international norm of behavior backed by enforceable treaties. Sometimes soft law becomes hard law over time as international norms first articulated in soft law come to be expected and relied on.

The UN works in a number of ways to promote and develop soft and hard law around the issues of sustainable development. It sponsors scientific research and publishes information critical to policy development. It also sponsors many conferences at which soft law principles of sustainability emerge, and consensus develops. It promotes the development of nongovernmental organizations with goals compatible with these principles. In addition, it offers member governments consulting services, work plans, and training.

The United Nations is an organization of member nations that is organized into five principle branches with many overlapping and redundant programs. The five main branches of the UN are the General Assembly, the Security Council, the Economic and Social Council, the UN Secretariat, and the International Court of Justice. Of these five branches, two have important responsibilities for sustainable development: the General Assembly and the Economic and Social Council. Within each branch, there are a variety of subgroups with different types of missions and goals, from those devoted to a single conference or treaty to those with multiple stakeholders, partnerships, and multifaceted responsibilities.

The most significant UN programs on sustainable development are authorized by the General Assembly and the Economic and Social Council (sometimes called ECOSOC). The General Assembly is composed of all the member nations and headed by the UN secretary general. Each member nation has one vote in the General Assembly, and they pass a type of legislation called resolutions that are nonbinding recommendations to member nations. The Economic and Social Council is a group of 54 member nations that collect information, coordinate policy development, and initiate programs related to the UN's economic, social, and development goals and objectives. The council elects its own president who serves for one year.

The UN has several programs important to sustainability in many different areas of its operations, including its programs on sustainable development, population, women, urbanization, environment, and children. Population programs are important to sustainability because as populations expand, so do human demands on the environment, although population growth in developed countries has a far greater environmental footprint than population growth in less developed nations. Population growth diminishes as women gain educational and employment opportunities. Opportunities for women often translate into opportunities for their children. Urbanization is proceeding at such a dramatic pace that by the year 2030, 60 percent of the world's population will be living in cities. There will be many more megacities, cities with populations of 10 million people or more, and megalopolises. Many of these areas will have high concentrations of poverty. Cities, however, offer the chance to manage waste and pollution efficiently because of their concentration, and to achieve diversified economies that can withstand economic cycles.

Examples of important sustainability-related programs of the General Assembly are the UN Environmental Programme or UNEP, United Nations Human Settlements Programme or UN–HABITAT, and the UN Development Programme or UNDP. All of these agencies are involved with external partners including other international organizations, development agencies, and nongovernmental organizations in the production of the Millennium Ecosystem Assessment and other initiatives.

United Nations Environmental Programme (UNEP)

UNEP coordinates all the environmental activities of the UN. It began at the United Nations Conference on the Human Environment in Stockholm in 1972. Many scientists and environmental activists were concerned that many environmental problems spanned the boundaries of nations and that rising world populations could make these problems worse in ways no one nation could effectively handle. After the conference the United Nations passed Resolution 2997 in December 1972, establishing the United Nations Environment Programme as a permanent institution charged with protection and improvement of the environment. Resolution 2997 also charges the UNEP with the missions of promoting international cooperation on the environment; reviewing global environmental issues so that governments give them adequate consideration; promoting the acquisition, assessment, and exchange of environmental knowledge; and reviewing the environmental impact of environmental policies on developing countries.

The UNEP is governed by a Governing Council of 58 members elected by the UN General Assembly for three-year terms. Member seats are allocated on a global regional basis. The headquarters is in Nairobi, Kenya, with six regional offices around the world. It has seven divisions: Early Warning and Assessment; Environmental Policy Implementation; Technology, Industry and Economics; Regional Cooperation; Environmental Law and Conventions; Global Environmental Facility Coordination; and Communication and Public Information. UNEP is a big player in major international environmental initiatives. It publishes

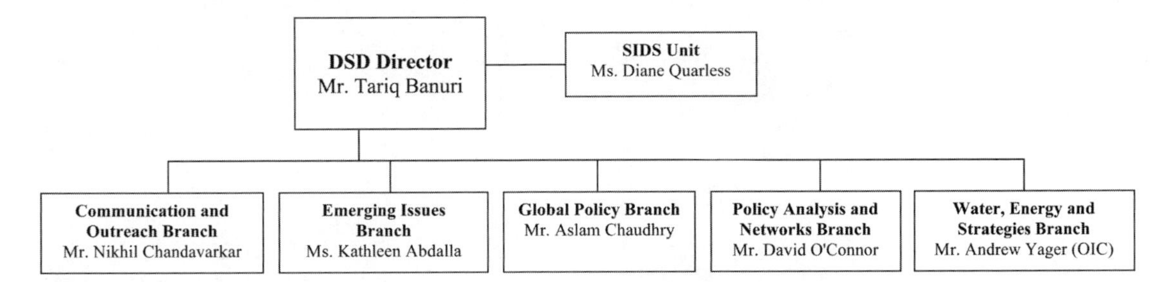

FIGURE 2.15 • UN Environment Programme Organigram.

many books and reports. UNEP's intermediate strategy for 2010–2013 is to prioritize climate change, disasters and conflicts, ecosystem management, environmental governance, harmful substances and hazardous wastes, and resource efficiency-sustainable consumption and production. For more information on UNEP see www.unep.org.

The United Nations General Assembly created the World Commission on Environment and Development (UNCED) in 1983. UNCED was renamed United Nations Environmental Programme (UNEP). Its first chair was Prime Minister Gro Harlem Brundtland of Norway. Its charge was to examine the global environment and development, assess critical environmental problems, develop proposals for solving them, and raise the international level of understanding of them. UNEP is responsible for several major programs related to sustainability. A few are listed here.

The World Commission on Environment and Development is most famous for laying the groundwork for modern views of sustainability. That view was expressed in their report, "Our Common Future: Report of the World Commission on Environment and Sustainability" issued March 20, 1987. The Report summary states:

> UNEP coordinated the Conference on Environment and Development known as the Earth Summit in Rio de Janiero in 1992. It also participates in the functions of the Intergovernmental Panel on Climate Change that won the Nobel Peace Prize in 2007.

United Nations Conferences on Environment and Development have become the forums in which these key concepts have been turned into implementable policy statements. The agreements and statements resulting from these conferences are often identified by their host city. Perhaps

FIGURE 2.16 • United Nations Conference on Environment and Development (UNCED) also known as the Rio Summit, Earth Summit. Opening ceremonies at the Earth Summit in Rio de Janeiro, Brazil, were held on Thursday, June 3, 1992. AP Photo/ Eduardo DiBaia.

the most famous of these conferences was the "Earth Summit" held Rio de Janiero in 1992. At this conference the nations of the world, including the United States, agreed to implement seven key concepts to ensure sustainable development in a declaration called the "Rio Declaration." They also wrote out a work plan called "Agenda 21," which remain the source of much international controversy to this day. Another famous agreement from this conference was the UN Framework Convention on Climate Change (UNFC3). A subsequent round of negotiations based on it was held in Kyoto, Japan, in 1997, resulting in the Kyoto protocols on climate change and the limits on emissions of greenhouse gases. Although a signatory to the protocols, the United States has not moved forward with ratification of the protocols, even though it is the world's largest producer of carbon dioxide, because it disagrees with the exemptions given to developing economies like China and India.

The United Nations Conferences on Environment and Development have become the forums in which key concepts of sustainable development have been turned into "hard" international law including implementable policy statements and multilateral treaties. These agreements and statements resulting from these conferences are often identified by their host city.

UN Commission on Sustainable Development (UNCSD)

In 1992, the UNCSD, which took over activities related to sustainable development.

The "Earth Summit" and the Rio Declaration (1992)

The United Nations hosted a Conference on Environment and Development in Rio de Janiero, Brazil, in 1992. The conference itself is often referred to as the Earth Summit. In a declaration called the "Rio Declaration," member nations including the United States, agreed to link future development to sustainable development principles. This conference produced a work plan for sustainable development that has guided all subsequent efforts to implement sustainability and sustainable development. This work plan is called "Agenda 21," anticipating the achievements of sustainable development in the 21st century.

Agenda 21

Agenda 21 is a work plan for sustainable development that has guided all subsequent efforts to implement sustainability and sustainable development. It adopted the core principle that the major cause of environmental degradation is unsustainable process of manufacturing, and unsustainable levels of consumption mainly in industrialized nations. Agenda 21 comprises 27 principal chapters. These remain the foundation of many governmental and nongovernmental efforts to implement sustainability at the local, national, and international level. The 27 chapters identify four key concepts for laws and public policies to achieve sustainability:

- *Integrated Decision Making*: Development decisions should to consider environment, economics, and equity at the same time and of equal importance in all development decisions.
- *Sustainable Consumption and Production*: Methods of production and products should be produced without waste or pollution.
- *Intergenerational Equity:* Development decisions should not compromise the ability of future generations to meet their own needs.
- *Precautionary Principle:* When harm to the environment or human health will result from development, reasonable measures should be taken to prevent harm even if the scientific evidence is inconclusive.

The precautionary principle offers guidance in decision making to prevent serious and irreversible harm to the public or the environment. It places responsibility on the proponent of the proposed activity to establish that their actions will not, or are unlikely to, result in significant harm. The burden of proof falls on those who would advocate taking the action, even when there is a lack of scientific consensus that harm would occur. The European Commission Communication on the Precautionary Principle notes: "The precautionary principle applies where scientific evidence is insufficient, inconclusive or uncertain and preliminary scientific evaluation indicates that there are reasonable grounds for concern that the potentially dangerous effects on the environment, human, animal or plant health may be inconsistent with the high level of protection chosen by the EU." (February 2, 2000) Other principles of sustainability generally include:

- *Polluter Pays:* The costs of pollution should be paid by the source of that pollution. Governments should make sure that these costs are not paid by others by implementing fines, enforcement, and taxes.
- *Public Participation:* The best development decisions are made by people who have to live with their consequences. Governments must include these people in the decisions that affect them by direct access to the decision-making processes.
- *Differentiated Leadership:* Sustainability can mean very different things depending on a country's economic and social conditions. In some countries, the need to develop is driven by the basic human needs, and in others consumption of resources far exceeds basic human needs. Sustainability may mean reduction in consumption for some, and increasing development without unsustainable methods of production in others. Leadership toward sustainability must mutually recognize different approaches for different conditions.

Potential Changes in the Burden of Proof for Chemical Safety and Risk Under Sustainability

The European Union (EU) Commission accepted a new program called REACH. Under this program chemicals are managed under the precautionary principle. A list of about 3,000 chemicals must be proven "safe" by the manufacturer of the chemical. In most countries the burden is on citizens to prove that the chemical caused them specific damage. If the damage is too great, they may not make it to a court trial or other official determination of cause. The REACH program changes this burden so that the manufacturer must prove that the chemical is safe. The manufacturer must go through an authorization process and may have to provide a list of safer alternatives. Final action from the EU Commission is pending.

Local Agenda 21

The UN encourages every local government to engage in a local process establishing its Agenda 21 goals and work plans. This is consistent with Chapter 28 of Agenda 21, the principle that states that changes toward sustainability should be achieved at the lowest level of government consistent with effectiveness. To that end, the UN, together with other partners such as the nonprofit organization International Council of Local Environmental Initiatives, has been assisting local governments to convene stakeholders and draw up local Agenda 21 plans of action.

References

Dernbach, John, ed. 1992. *Stumbling toward Sustainability*. Washington DC: Environmental Law Institute, pp. 45–63.

International Council for Local Environmental Initiatives (ICLEI) and Local Agenda 21

Local governments are made up of many different kinds of systems and organizations, from cities and towns to counties and state governments. They are often where the rubber meets the road in terms of implementing any changes. ICLEI is an international organization of governments with powers that direct communities toward sustainability. This organization has made sustainability a part of its programs and has created a toolkit and approach to localizing the mandates of sustainability and Agenda 21. This approach includes identifying local barriers to sustainability, reducing the impact that their localities have on natural resources such as fresh water and cultivatable land, and developing methods that make these measures transparent and accountable to their local stakeholders.

To date, more than 6,400 communities in 113 countries have participated in this program and have adopted local work plans for achieving sustainability.

UNFCCC and the Intergovernmental Panel on Climate Change (IPCC)

The UN Framework Convention on Climate Change was also a creation of the Earth Summit. It established limits on greenhouse gas emissions for all consenting nations, and it created the IPCC to support the work of limiting greenhouse gas emissions. The IPCC began when the UNEP and the World Meteorological Organization met in 1988 to review the entire scientific and technical peer reviewed literature on global climate change. The IPCC dos not perform any original scientific research. Member countries come from UNEP or the World Meteorological Organization and they choose the scientists who participate. These scientists are considered the leading experts on climate change and their reports are relied on heavily in all world pronouncements on the issue. Many nations base their environmental policies and programs on their reports. Most of their reports try to make it assessable for policymakers. The reports generally include discussion of future development scenarios relating to social, economic, technological, and energy use. They specifically include analyses of future greenhouse gas emissions.

Currently, the IPCC has three Working Groups and the Task Force on National Greenhouse Gas Inventories. The mission of Working Group One is to study the research on the physical science basis of climate change. The mission of Working Group Two is to handle the research on climate change impact, and vulnerability. Working Group Three is to handle mitigation of climate change. All groups face powerful methodological and data challenges.

The IPCC has been criticized for being too cautious in its scientific approach to climate change. Part of the scientific methods is to count only those changes that fall within prescribed scientific levels. Some ICC reports limit projections of temperature changes to those that fell within a 90 percent confidence level. This discounted a small probability that larger scale unpredictable changes, such as methane gas releases from the Arctic tundra or complete loss of the Antarctica and Greenland ice field. The loss of the reflective ice (called the albedo effect) and the absorption of the sun's heat by the then exposed water could dramatically increase global warming. In terms of sustainability, some of the scientific assumptions of the IPCC could irreparably harm the systems of life on which future generations depend. They assume that benefits for future generations should be discounted relative to the costs on present generations. They also make little consideration for the impacts on the most vulnerable nations, either in terms of the small probability of nonlinear, rapid environmental changes or in the impacts of an intergenerational cost-benefit analysis. The last set of criticisms relate to how the scientific literature is developed. It is from research that is technical in nature and that is peer reviewed. It is not from people who may be experiencing the environmental changes first hand, as is the case with many indigenous people. Much of the research comes

FIGURE 2.17 • Nobel Peace Prize winners Al Gore, center, and Chairman of the Intergovernmental Panel on Climate Change Dr. Rajendra K. Pachauri, right, receive their medal and diploma from Nobel Committee Chairman Ole Danbolt Mjoes, left, at City Hall in Oslo, December 10, 2007. AP Photo/John McConnico.

from studies performed by those with a stake in the profit made from a given natural resource, such as timber, minerals, and water. Further, the nature of peer review limits the publication of a given work to a traditional paradigm, or way of thinking. Peers are often those of a singular discipline, so that a given piece of environmental or ecological research that is outside the usual disciplinary boundaries will be rejected. In this way, the IPCC is considered by some to be self-limiting and may be underpredicting the potential for climate changes. In response to some of these criticisms, the IPCC issued a decision framework prioritizing assessment reports and other measures to protect the credibility of the organization in 2008.

The Kyoto Protocol (1997)

A round of negotiations to establish national limits on greenhouse gas emissions was held in Kyoto, Japan, in 1997, resulting in the Kyoto protocols on climate change. The United States has not ratified the Kyoto Protocol, although the majority of the other nations have done so. The Kyoto Protocol is the leading international agreement for affecting human-caused climate change. Most of the other international climate change studies and policies feed into the Kyoto Protocols. These protocols require honest and accurate assessments and policies about greenhouse gas emissions. They require strict limits on what developed countries are allowed to emit. They also require developed countries to assist developing countries in ways that limit greenhouse gas emissions. The Kyoto Protocol are the first significant international step to meaningfully affect global natural systems. As such, their future impact on sustainability is immense. It is a new and dynamic set of principles and agreements, and will be revisited many times in the future.

The Millennium Ecosystem Assessment (2000–present)

The Millennium Ecosystem Assessment (MA) is, to date, the most comprehensive survey of the ecological state of the planet. The MA was called for by the then-secretary general of the United Nations, Kofi Annan, in 2000, and began its work in 2001. The MA was undertaken by an international network of scientists and other experts. More than 1,300 authors from 95 countries were involved in the MA and were divided into four working groups. Three of the working groups, Condition and Trends, Scenarios, and Responses, focused on global assessment goals, whereas the fourth working group focused on subglobal assessments. The resulting assessment was divided into four technical volumes that were reviewed by experts and governments, and more than 600 individual reviewers worldwide provided around 18,000 individual comments. The assessment lasted four years and concluded in 2005 with its report on the consequences of ecosystem change for human well-being and the scientific basis for actions needed to enhance the conservation and sustainable uses of the Earth's resources. The MA had four main findings:

1. Over the past 50 years, humans have changed ecosystems more rapidly and extensively than in any comparable period of time in human history, largely to meet rapidly growing demands for food, fresh water, timber, fiber, and fuel. This has resulted in a substantial and largely irreversible loss in the diversity of life on Earth.

2. The changes that have been made to ecosystems have contributed to substantial net gains in human well-being and economic development, but these gains have been achieved at growing costs in the form of the degradation of many ecosystem services, increased risks of nonlinear changes, and the exacerbation of poverty for some groups of people. These problems, unless addressed, will substantially diminish the benefits that future generations obtain from ecosystems.

3. The degradation of ecosystem services could grow significantly worse during the first half of this century and is a barrier to achieving the Millennium Development Goals.

4. The challenge of reversing the degradation of ecosystem while meeting increasing demands for services can be partially met under some scenarios considered by the MA, but will involve significant changes in policies, institutions, and practices that are not currently underway. Many options exist to conserve or enhance specific ecosystem services in ways that reduce negative tradeoffs or that provide positive synergies with other ecosystem services.

The findings basically outlined the negative impact of human actions on the Earth's natural capital. The MA showed that human's cannot take the Earth's ability to sustain future generations for granted at this rate of environmental degradation; however, if appropriate substantial action were to be taken, it would be possible to reverse some degradation over the next 50 years. The MA is an assessment of data that was already available; it is unique in that it was a global assessment that presented a consensus view of the current state of the planet. Consensus is one of the cornerstones of change. Furthermore, the MA also identified a number of "emergent" findings or conclusions that were the result of examining a large amount of information together.

One of the simultaneous strengths and weakness of the MA is the gaps in knowledge about the status of the planet's ecosystem. There is relatively limited information available at a local and national level about the status of ecosystem services and even less information about the value of nonmarketable services. There is also limited information about the economic costs of ecosystem degradation. In an increasingly market-based world, it is imperative for economies to have information and assessment of economic loss of human action. The MA created awareness of these gaps in knowledge and stimulated data gathering and assessment to eliminate the gaps.

The overall aims of the MA were to contribute to improved decision making concerning ecosystem management and human well-being. The MA work groups expect that there will be significant adoption of the MA conceptual framework that will continue to help meet the planet's sustainability needs and reverse degradation. The MA indicated that changes be instituted firmly and quickly. It was recognized that, as humanity has the power and ability to prevent the damages to the

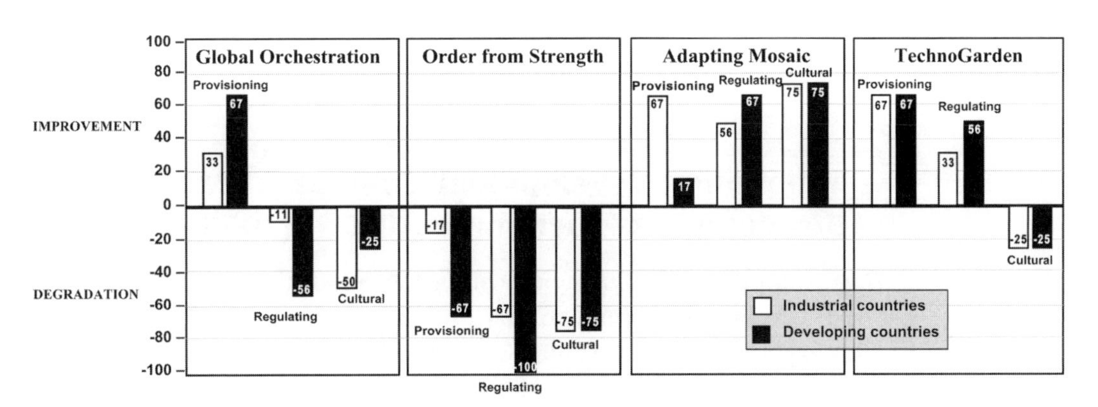

FIGURE 2.18 • Number of ecosystem services enhanced or degraded by 2050. 100 percent degradation means that all the services in the category were degraded in 2050 compared with 2000, whereas 50 percent improvement could mean that three of six services were enhanced, and the rest were unchanged, or that four of six were enhanced and one was degraded. The total number of services evaluated for each category was six provisioning services, nine regulating services, and five cultural services. Philippe Rekacewicz, Emmanuelle Bournay, UNEP/GRID-Arendal • http://maps.grida.no/go/graphic/number-of-ecosystem-services-enhanced-or-degraded-by-2050.

planet, it is also our duty to do so. One of the most important issues brought up was the effects of environmental damage to the underdeveloped and poor people of the world. The report urged the nations of the world to work harder to achieve a sustainable future.

Millennium Ecosystem Assessment: Businesses and Industries

The MA organization initially published a series of synthesis reports about how the ecosystem findings will affect human health and well-being. The synthesis reports focus on biodiversity, desertification, business and industry, wetlands and water, and health.

The business and industry report focused attention on the risks that degraded ecosystems create for businesses and industries, as well as the opportunities for strategic long-term thinking in ways that would stimulate sustainable behaviors currently. This synthesis report on business and industry pointed to three major consequences:

1. If current trends continue, ecosystem services that are freely available today will cease to be available or become more costly in the near future. Once internalized by primary industries, additional costs that result will be passed downstream to secondary and tertiary industries and will transform the operating environment of all businesses.
2. Loss of ecosystem services will also affect the framework conditions within which businesses operate, influencing customer preferences, stockholder expectations, regulatory regimes, governmental policies, employee well-being, and the availability of finance and insurance.
3. New business opportunities will emerge as demand grows for more efficient or different ways to use ecosystem services for mitigating impacts or to track or trade services.

Millennium Development Goals and UN Development Programme (UNDP) (2000–2015)

In 2000, the General Assembly of the UN adopted a declaration of goals for development to end poverty by 2015.

The Millennium Campaign adopted eight goals:

1. End hunger
2. Universal primary education
3. Gender equity
4. Child health
5. Maternal health

6. Combat HIV/AIDS
7. Environmental sustainability
8. Global partnership

The UN is involved in supporting development through the UNDP, a significant partner in implementing the Millennium Development Goals. Other major partners in UNDP projects include The World Bank and International Monetary Fund, and the World Health Organization. UNDP publishes an annual index of developmental measures including poverty, literacy, education, life expectancy, and other factors called the Human Development Index or HDI.

The Green Economy Initiative: Environmental Sustainability and Job Creation (2008)

Beginning in 2007, widespread global market failures signaled global economic depression triggered by speculation, lack of management, and oversight of natural capital resources. The global environmental and ecological failures triggered by the same types of human behavior were well documented in the Millennium Ecosystem Assessments (see Government Involvement). In response to this global economic crisis, UNEP has called for a redesign of the global economy to invest in clean technologies and environmental sustainability of our natural resources in its Green Economy Initiative. Some have called this initiative a global Green New Deal.

Three "pillars" shape this policy initiative. First is the establishment of national and international natural services accounts. These accounts attempt to identify a financial value for the assets that nature provides, as urged by green economists. Second, the initiative calls for encouraging green job creation by using the power of government to establish public policies that encourage such work. Last, the initiative calls for governments to establish instruments and market signals that will accelerate economic transition. UNEP will create a comprehensive assessment and toolkit, including ideas for shifting subsidies and using other market-based mechanisms for change.

Five industrial sectors are the focuses of this effort. They are clean energy and clean technologies, rural energy, sustainable agriculture, ecosystem infrastructure, forests, and cities.

THE WORLD TRADE ORGANIZATION (FORMERLY GATT)

The World Trade Organization (WTO) is an association of member nations who have agreed to eliminate trade barriers between themselves. It is the largest free trade agreement in the world. It was known as the General Agreement on Tariffs and Trade (GATT) until 1995. The process of negotiations about different types of trade barriers is a continuing process. Negotiations are known as "Rounds" and each Round is called by the name of the city in which it took place. Free trade agreements between nations have increased dramatically since World War II. Under these agreements, member nations agree to eliminate trade barriers between each other. A typical trade barrier is a tariff or customs duty, types of taxes imposed on imported goods. Another type of barrier is a subsidy paid in one country that makes the market price of the subsidized product cheaper without regard to its actual costs of production. Subsidies can have the effect of making one nation's products seem more or less expensive than goods of another competing nation.

The WTO and other free trade organizations have been instrumental in expanding markets globally. These organizations and their agreements have increased the size of markets, together with the size of businesses that compete globally, and the production, transportation, and use of consumer goods globally. This success has also been accompanied with the greatest increase in damage to natural ecosystems and climate changes in all of human history. In addition, these markets have increased the difference between rich and poor people and communities, even in developed nations like the United States, instead of creating a broad middle class. The challenge of sustainability, and the question for globalized markets, is whether the development power of globalized trade can be harnessed so that it does not degrade the environment or increase global poverty.

The WTO itself allows member countries to enact measures necessary to protect human health, animal or plant life, or conservation of exhaustible natural resources even when those national measures impose barriers to trade with other member nations. These measures are called sanitary and phytosanitary measures. In this way, member nations may infuse their substantive values such as caring for the environment and improving conditions of poverty into the trade mechanisms of the WTO; however, WTO imposes limits on the ability of members to enact such national measures. If WTO concludes that these measures are arbitrary or unjustifiable discrimination against other members, it may direct the offending member nation to bring its national laws into conformity with WTO principles. Occasionally, a losing member may resist or delay in establishing conformity, and the WTO will allow retaliatory trade measures by the winning party.

WTO disputes are initially reviewed by a three-person panel of trade experts. Its report is binding unless it is appealed to the WTO membership composed of all member nations. WTO also has an Appellate Body that reviews disputes on legal issues that panels have identified. The Appellate Body issues decisions on these legal matters, and their decisions are final unless rejected by the WTO membership.

Several interesting trade disputes have involved claims that measures related to environment or equity were not justifiable. These disputes involved trade barriers enacted by one member nations against imports from another member nation of genetically modified foods that had been banned from import into Europe—beef containing artificial hormones, and tuna caught by unsustainable methods.

THE NORTH AMERICAN FREE TRADE AGREEMENT (NAFTA)

The United States, Mexico, and Canada have joined together to form a trade association whose goal is to allow movement of goods and services across their mutual borders without restrictions like tariffs. National laws regarding the protection of the environment and fair labor standards differed substantially between the three signatory nations. U.S. environmentalists and unions opposed the agreement because of fears that U.S. norms and standards would become unenforceable as trade barriers because they increased the cost of products and production methods subject to U.S. rules on environment and labor protection. As a result of these concerns, additional agreements allowing for the protection of environmental and labor conditions in the United States were signed between Mexico and the United States. The so-called side agreements did nothing to alter national standards in either the United States or Mexico.

NAFTA and free trade principles have not eliminated controversies about trade between the nations of NAFTA. These continuing controversies include concerns about water resources from Canada being traded as commodities, Mexican immigration, and the consequences of NAFTA for jobs in the United States. In terms of regional decision making about matters of sustainability, NAFTA provides North America with a geographically coherent basis for making important decision regarding its ecosystems, economy, and healthy communities. This basis or platform for decision making does not automatically engage issues of sustainability without intentional leadership and action within member nations.

UNITED STATES

Obama Initiatives: Clean Energy, Climate Change, and Green Jobs

Elected in November 2008, and inaugurated in January 2009, Barack Hussein Obama, the 44th president of the United States began his administration with sharp changes from the preceding administration of George W. Bush on topics related to sustainability. Almost immediately, the Obama administration reversed a Bush administration decision refusing to allow the state of California to demand greater fuel efficiency standards of its automotive fleet than the federal standards. This reversal had a significant impact on greenhouse gas emissions, and climate change because California is one of the top 10 economies of the world, and is responsible for the production of an estimated 1.4 percent of greenhouse gases in the world today.

In another early action, the Obama administration repealed a Bush administration policy prohibiting the use of federal funds for abortion-related services abroad. Internationally, these services have a major impact on the ability of women to control the size of their families, and, with that control, gain opportunities for education and employment for themselves and their children. U.S. official direct assistance had been unable to fund any abortion-related services in foreign countries under the Bush administration. These opportunities are instrumental in lifting women, children, and their communities out of poverty.

President Obama's economic recovery plan also focuses on sustainable energy and "green" jobs for major investments.

Environmental Assessments

The National Environmental Protection Act (NEPA) enacted in 1970 established a process by which environmental effects of action can be identified and considered in advance of decision making. This process is required for federal action, but it is a model that can be adapted for use by other organizations to consider the risks and opportunities of their actions on the environment in strategic planning. Businesses may choose voluntarily to behave with precaution, taking reasonable measures to protect human health and the environment before harm is done, and before harm is established as a scientific probability. Such proactive, preventive action is the basis for sustainable, environmentally and socially responsible behavior.

The Millennium Ecosystem Assessment also encourages businesses to evaluate risks and opportunities to behave with precaution in the accompanying synthesis report interpreting the results of the ecosystem assessment for businesses.

(See Ecosystems and Human Well Being Opportunities and Challenges for Businesses and Industry A Report of the Millennium Ecosystem Assessment, Washington DC World Resources Institute 2005.)

NEPA is one method for considering environmental consequences as well as social and cultural consequences of action. Most of sustainability policy hinges on environmental assessments of the human impacts on the environment. It has been assessments of the environment that warn of the degradation of natural systems. It is environmental assessment policies that determine if and when and what kind of environmental mitigation is necessary. Most environmental assessments relied on by governments are based on some degree of science. There are international standards of environmental impact assessment such as those put forward by the United Nations Social, Educational, Social, and Cultural Organization.

On June 27, 1985, the European Council of Ministers adopted a rule that required its members to adopt environmental assessment procedures. Its members include Austria, Belgium, Denmark, Finland, France, Germany, Greece, Ireland, Italy, Luxembourg, The Netherlands, Portugal, Spain, Sweden, and the United Kingdom. They include a threshold determination of whether the environmental impacts are significant, and are applicable to both government and private projects. They have a list of projects that require an environmental impact statement (EIS) and another list of projects for which an EIS is discretionary.

Communities do their own environmental assessments, and citizen monitoring of the environment is on the increase. Industries also do their own environmental assessments. There are many kinds of environmental assessments. Models of environmental assessment that ignore, diminish, or underreport actual environmental impacts are inadequate for sustainability purposes. The purpose of a given environmental assessment is very important. It is likely that many of the currently used environmental assessment models will be the springboard for sustainability environmental assessment models. ***See also* Volume 2, Chapter 2:Language of Sustainable Business.**

The U.S. Model

The first U.S. federal law requiring environmental impact assessments was NEPA. Many states have since developed their own NEPAs, and some tribes have developed their own Tribal Environmental Policy Acts. An overall critique of the environmental assessment policy context is that it does not cover enough of those human activities that result in environmental impacts. It often fails to measure cumulative impacts or ecosystem impacts over long periods, for example. Some environmentalists want all environmental impacts covered by new policies such as a municipal environmental impact statement. Pushing the NEPA environmental assessment model to cover all environmental impacts that threaten natural systems on which all life depends would enrich it to sustainability levels.

NEPA requires a detailed environmental impact statement for major federal actions significantly affecting the quality of the human environment. The express legislative purpose of NEPA is:

> To declare a national policy which will encourage productive and enjoyable harmony between man and his environment; to promote efforts which will prevent or eliminate damage to the environment and biosphere and stimulate the health and welfare of man; to enrich the understanding of ecological systems and natural resources important to the Nation; and to establish a Council on Environmental Quality. NEPA section 2.

The following underlying purposes of NEPA are much in line with current principles of sustainability:

1. To fulfill the responsibilities of each generation as trustee of the environment for succeeding generations
2. To assure for all Americans safe, healthful, productive, and esthetically and culturally pleasing surroundings
3. To attain the widest range of beneficial uses of the environment with degradation, risk to health or safety, or other undesirable and unintended consequence
4. To preserve important historic, cultural, and natural aspects of our national heritage and maintain, wherever possible, an environment that supports diversity, and variety of individual choice
5. To achieve a balance between population and resource use that will permit high standards of living and a wide sharing of life's amenities
6. To enhance the quality of renewable resources and approach maximum attainable recycling of depletable resources.

NEPA, section 101(b)

A fundamental weakness of NEPA is that the environmental assessment is not a document that determines policy; it is ultimately advisory only. This would make it unsuitable for purposes of sustainability because it would not be action oriented enough. Much of the purpose and goals of NEPA, however, follow many principles of sustainability. The processes and participants, or stakeholders, to the NEPA process would be the first used for sustainability environmental impact assessments.

The purpose of the NEPA environmental assessment is to reduce environmental impacts on the environment when possible. Economic considerations specifically and legally drive the decision-making process, however, because they are the overriding value in U.S. society. Nonetheless, the NEPA EIS process offers valuable ways to garner important environmental information.

There are four main sets of participants to the NEPA process. The first is the lead agency, which is the agency responsible for EIS preparation and for making the decision on the proposed action. The second set of participants is the EIS team. This group can differ widely from project to project, depending on the range of issues. Generally, it is an interdisciplinary group of specialists making scientific observations and decisions around these observations. They are usually scientists, engineers, and institutional planners. Each group member is supposed to be fair and unbiased. Each is supposed to examine each area as thoroughly as the level of significance of the environmental impact dictates. Group members can be employees of the lead agency or private consultants working under the lead agency, as well as agency staff.

Some advocates of greater inclusion in the environmental decision-making process and as a principle of sustainability criticize the NEPA process now because there is no citizen or community or environmentalist involvement in the core EIS team. The lack of inclusion, they claim, can cause them to issue a finding of no significant environmental impacts when, in fact, there are significant environmental impacts. The level of significance of environmental impacts is very controversial under NEPA. Short-term impacts resulting from most construction techniques are not considered significant. The environmental impacts of the mitigation techniques themselves are not considered. Some industry stakeholders argue that simple compliance with air, water, and land environmental laws should be considered part of the mitigation package. Under a regime of sustainability policy, significance would be tied to potential for irreparable damage to natural systems on which future life depends. There would be little to mitigate if the damage were irreparable, but controversies ensue until the evidence almost reaches levels of species extinction, as in overfishing.

The third group that is part of the U.S. NEPA process is the project proponent. This is usually a private developer or landowner, or sometimes an agency. It can also be called an applicant or a sponsor. If project proponents need money, environmental permits, or governmental approval, they probably have to begin the environmental impact process. They have to provide accurate and complete information about the design, construction, and operation of the proposal. They are supposed to share all drawings, feasibility studies, environmental information, and building designs with the lead agency. The lead agency can request more information or explanations of the information provided. Project proponents often complain about the intrusion into business practices and the amount of time an EIS involves. Many claim it scares away potential investors. Sometimes the process of doing an EIS uncovers previously unknown legal liabilities, such as an illegal hazardous waste site. This also affects investors' perceptions of the risk involved with the project. If there are significant environmental impacts that have to be mitigated, the cost of mitigation could be expensive. These contingent liabilities of time-consuming, unknown environmental liabilities, and cost of mitigation also affect the project proponent, as well as the NEPA environmental assessment process.

The role of the public as a participant is a varied and growing one under U.S. NEPA. Generally interested parties of the public are allowed to provide input to the lead agency in the scoping process to narrow the issues and alternatives in a draft EIS, and to review the draft EIS. Many communities have felt excluded by this process. The lead agency generally selects which parts of the public are allowed to participate. Communities want more of a say in developing project alternatives and in designing mitigation schemes. Under NEPA, public review participants include private citizens, Indian tribes, other agencies that have expertise or jurisdiction, and interested parties who have requested notification.

The NEPA Environmental Assessment Process

The process begins when project proponents submit their application to a federal agency. At this stage of the process, the application may not indicate all the environmental impacts but show preliminary designs and concepts. If it is a state environmental policy act (SEPA), private or local agency actions may not necessarily be covered. Some states require only lead agencies to submit EISs.

After the proposal is submitted, the lead federal agency determines whether the project is categorically excluded from EIS requirements or if it is exempt from them. Both the federal agencies and state governments have these categories. The policy justification for these categorical exemptions is that these are activities that do not usually cause significant impacts and therefore would not require an EIS. Many environmentalists and sustainability proponents, however, argue that this is not always the case. For example, Community Development Block Grants were largely categorically excluded from federal EIS requirements under NEPA. These funds went to many programs in urban areas that had direct and indirect environmental impacts that were not counted or assessed. An adequate platform for a sustainability policy would need to count environmental activities without categorical exclusions or exemptions.

The next step in the U.S. NEPA process is to make a threshold determination of whether there are significant impacts to the environment because of the project. This is a highly controversial and litigated area of the law. Threats to endangered species, wetlands, historic and cultural areas, and controversy itself can trigger a level of significance that requires the EIS. Issues of environmental injustice and racial disproportionality can be a significant impact. If the lead agency is uncertain about whether there will be significant impacts on the environment, it performs an Environmental Assessment, or EA to determine the potential for significant environmental impacts.

There is substantial community concern and sustainability criticism for this step of the process. Communities are not given notice of this EA or even the project application. They have no opportunity for involvement to say what they think the environmental impacts of the proposal would be to them and their environment. The EA is often limited to long-term, direct impacts on the study site alone. Some claim the study area is too small to measure ecosystem impacts in most projects, and that the environmental impact study area is manipulated to decrease environmental liability and significance of environmental impacts. For example, in a site where a federal courthouse was proposed, a leaking underground storage tank from an abandoned gasoline station was found. If it were included in the study area, it would have shown a plume of petrochemical pollution from the leaking underground storage tank through the soil, to the water table and to a nearby river. The cost of cleanup and mitigation of this hazard would be expensive and time consuming. The study area was redefined to exclude the abandoned leaking underground storage sites. Although the community resisted because it wanted to get the site cleaned up, the federal government, and its EIS process, expressly preempts state and local environmental laws.

If the lead agency finds no significant environmental impacts, it issues a Finding of No Significant Impacts, or FONSI. This is often the first notice the community receives about a project in its midst. Some communities wholeheartedly endorse any economic development despite environmental consequences. Other communities express shock and outrage at their lack of inclusion on the threshold issue of significant impacts. Environmentally burdened communities are especially sensitive to late notice and exclusion from decisions that directly affect them. Many environmentalists express concern that a study that is more thorough was not performed and that ecosystem and cumulative effects were not included. Sustainability proponents find this stage of the process lacking because of lack of inclusion and because of lack of ecosystem- or biome-based study. The U.S. Environmental Protection Agency's Council on Environmental Quality defined ecosystem as:

> An ecosystem is an interconnected community of living things, including humans, and the physical environment with which they interact. Ecosystem management is an approach to restoring and sustaining health ecosystems and their functions and values. It is based on a collaboratively developed version of desired future ecosystem conditions that integrates ecological, economic, and social factors affecting a management unit defined by ecological, not political, boundaries.
>
> The Twenty-fourth Annual Report of the Council on Environmental Quality, 1993. (available on line at ceq.hss.doe.gov/nepa/reports)

If the lead agency does find significant environmental impacts, then under the U.S. NEPA process they issue a Notice of Intent (NOI) to prepare an EIS statement. The NOI is published in the *Federal Register.* This publication comes out daily and is issued to federal depository library institutions. It constitutes public notice of agency actions such as rules, regulations, and EISs.

From here, the public scoping process begins. Scoping is an important part of the U.S. NEPA process because it can determine the actions, alternatives, environmental effects, and sometimes mitigation measures included in the EIS. Different federal agencies involve the public to different degrees in the scoping process. A common complaint, however, is that only those members of the public that agree with the proposal are allowed to participate in a meaningful manner. Many environmentalists and most sustainability proponents would consider the actual scope of the EIS to be too narrow and small to be applied to environmentally based sustainability approaches. Agencies have the discretion to choose members of the public. Until communities complained about environmentally unjust disproportionate environmental impacts, few members of the public were engaged in the scoping process. Scoping is done informally and formally by the lead agency, often in close consultation with the project proponent or their consultants. Communities and tribes have felt excluded from the process because they *were* excluded. The lead agency determines the size and scale of the environmentally affected area. This is a controversial decision because it underestimates environmental impacts according to many environmentalists.

Another important area of public involvement under the U.S. NEPA process is public review of the draft EIS. The draft EIS contains all the alternatives to significant environmental impacts. The range of alternatives, including the no action alternative, is often very small. The more alternatives considered, the more expensive and time consuming the EIS process can be. The draft EIS can contain important environmental information that could be useful as baseline information in later sustainability assessments.

Public review of the draft EIS is limited under U.S. NEPA processes. Only comments within a period that address certain questions posed by the lead agency are reviewed. Agencies allow little public review from groups that simply do not want the project at all because of environmental impacts. The public is composed of developers, special interest groups, environmental and economic advocacy groups, individual citizens, and other reviewing agencies. Lack of inclusion at this stage of the NEPA process changed in the late 1990s, primarily because of the political pressure of environmental justice groups and the legal advocacy of environmental groups seeking information from ongoing EIS processes.

After the public review of the draft EIS, the lead agency reviews all the comments relevant to the significant environmental impacts and alternatives presented in the draft EIS. These are generally published in the *Federal Register*.

References

Bass, Ronald E. et al. 2001. *The NEPA Book: A Step by Step Guide on How to Comply with the National Environmental Policy Act.* Point Arena, CA: Solano Press.

Eccleson, Charles H. 2008. *NEPA and Environmental Planning: Tools and Techniques and Approaches for Practitioners.* Boca Raton, FL: CRC Press.

EIS: A Done Deal?

Environmental lawyers must generally wait until an administrative agency makes a final decision. This is a legal doctrine called "the exhaustion of administrative remedies." The purpose of the doctrine is to leave specialized and complex areas within the expertise of the administrative agency until the complainant has pursued all administrative avenues, giving the administrative agency an opportunity to self-correct. This can be a long and expensive process, effectively excluding most poor and working people. Interagency appeals processes can take years. The final decision in many cases is not always clear. In the NEPA EIS process the final lead agency decision is called the Record of Decision and is published in the *Federal Register*. This is long after the EIS is complete. Many environmentalists and communities feel excluded from meaningful participation, leaving their interests unaddressed. Often their interests are in line with sustainability values and approaches, such as preservation of the environment for future generations and the application of the precautionary principle.

Economic Value Prioritized in U.S. NEPA

It is clear that the U.S. NEPA EIS process is laden with a strong economic value directed toward growth. It is unusual for a project to be denied. Many claim that the environmental mitigation measures claimed in the EIS are unenforceable. The whole decision is not mandatory, merely advisory. From a sustainability perspective, the lack of meaningful participation around ecosystem issues poses a challenge. Although the U.S. EIS process is supposed to examine cumulative impacts, and cumulative impacts are supposed to be a significant impact on the environment, in reality, they are ignored because of the time and resources necessary to evaluate and then mitigate them. The increased environmental scrutiny under assessment procedures, however, may dissuade projects with overwhelming harmful effects from even submitting an application. This is not necessarily the case because of the pro-growth assumption of the U.S. EIS process. Nuclear reactors and nuclear waste sites are subjected to stringent EIS procedures in most cases but are eventually given permission to operate, for example.

Another check on the U.S. NEPA process is the amount of time involved. An EIS can take a long and uncertain amount of time. From the project proponent's perspective, this can decrease investment and ultimately profitability. There is always a certain amount of pressure to streamline the process. This often occurs at the expense of public participation. In contested environmental issues, such as timber sales in national forests, industry feels that some public participation is done to help environmental lawyers prepare for their lawsuits once the administrative decision is final.

U.S. Environmental Protection Agency: Checks and Balances?

Most federal agency EIS under NEPA must clear the U.S. EPA, which ensures that the EIS is up to minimal standards. This occurs in two stages of the EIS process. The first stage is when the EPA evaluates the adequacy of the Draft Environmental Impact Statement. It places into one of three categories. The first category is "adequate." The standard for adequate is that the draft EIS sufficiently lists and describes the impacts of the alternatives developed, and that no further analysis is necessary. The next category is "insufficient information." This means that the draft EIS did not contain enough information for the EPA to assess environmental impacts that need to be avoided. It can also mean that new alternatives were identified that would reduce environmental impacts more than those considered. These new alternatives can come from many sources at this juncture in the process. Communities with late notice but strong political power can be one source. Other federal agencies can be another source. It is the decision of the lead agency to decide the scope of alternatives, but the EPA can label it insufficient if it finds an alternative that should be there. This then requires that the final EIS consider the alternative. It does not require that it accept it. The last category of EPA evaluation of the draft EIS is "inadequate." This means the draft EIS does address significant environmental impacts. It can also mean that new alternatives were introduced that would reduce environmental impacts but were outside the scope of alternatives available in the draft EIS. It requires that the draft EIS be revised and resubmitted for public review as a Supplemental Draft EIS.

The EPA can also evaluate the environmental impacts of the action in the draft EIS. There are four categories of evaluation. The first category is "lack of objections." The second category is "environmental concerns." This means that the EPA identified environmental impacts that should be avoided. Environmentalists have criticized the EPA for not using this category as a basis for broader environmental concerns, such as sustainability. With environmental concerns, the EPA indicates that they volunteer to work with the lead agency to mitigate the identified environmental concerns. The lead agency does not have to do so. The third category is "environmental objections." Here the EPA identified significant environmental impacts that must be avoided. In this case, the EPA intends to work with the lead agency to avoid these identified environmental impacts. The last category is "environmentally unsatisfactory." This means that the EPA identified very serious environmental impacts that could endanger the public health and environmental quality. These environmental impacts must be avoided and the final EIS must show it. If the final EIS does adequately deal with these environmental impacts, the matter is referred to the Council on Environmental Quality.

Real Property Environmental Assessments

The purpose of this type of environmental assessment is to avoid liability for past and present cleanup responsibilities on the site. The cost of clean land is a large and dynamic factor in a real estate transaction in many industrialized urban areas. A complex and controversial environmental policy question is whether an institution that finances the purchase of a site requiring cleanup is also liable for cleanup costs. Potential buyers and lenders interview contiguous neighbors, sample soil and water, and search public land and environmental records. In the United States, some banks do extensive environmental agency research at the federal, state, and local level. If the land is found to be contaminated, a whole other level of more probing, on-site assessments take place. These are not human health, ecological, or cumulative risk assessments. They are assessments of cleanup liability and generally assume low levels of cleanup limited only to that site. They may assess whether they can divide the property to avoid the contaminated portion of the site. Generally, at this level an environmental assessment must include the costs of remediation in the assessment. A traditional market value appraisal of real property does not offer much useful environmental information about the land. An environmental assessment does provide this information by including actual environmental condition of sites and the cost of cleaning them to the lowest possible standard, without assessment of biological or chemical risks.

Environmental Assessments: Are They Enough for Sustainability?

The state of environmental assessment is dynamic and growing around the world. The Council of European States requires a post-EIS phase to follow through on the state of any environmental impacts and to see if the promised mitigation is working to mitigate the environmental impacts. Their EIS process includes a research and development component so that it may learn lessons from the site and see how that site fights into its ecosystem. Soon technology will allow us to monitor every site on Earth. This allows monitoring of remote locations, such as the poles. It also allows for a broad regulatory potential for the environment, something needed for new sustainable policies. Environmental enforcement models rely heavily on assisting compliance. In the United States, if an environmental wrongdoer simply admits to the charge, the fine is reduced 50 to 70 percent, and most fines are not collected. Sustainability will require strong enforcement models. Strong enforcement models will be greatly aided by complete, real-time monitoring of the environment. These enforcement mechanisms will be to deter environmental behavior that degrades the environment. Technological improvements in environmental monitoring via satellites, cameras, and monitoring stations (staffed and unstaffed) may set the stage for sustainability policies. These improvements vastly increase the number of observations of natural systems, and that knowledge contributes to a burgeoning understanding of the interrelatedness of global ecosystems. Environmental assessments are asked to do more and more, such as include cumulative impacts analysis and ecosystem risk analyses. Emerging from the growing knowledge base available for environmental assessments, the growing application of environmental assessments to more projects, and corporate reporting of the triple bottom line (social, economic, environmental; see discussion in volume on Business and Economics) is a new emerging "sustainability" assessment.

The amount of growth of environmental knowledge is rapidly increasing, spurred by the knowledge that natural systems have limitations and that humans have a large effect on these natural systems. This rapidly increasing knowledge base of environmental limitations relies heavily on environmental assessment. That assessment, however, depends on the purpose for which it is used.

Some industry and corporate sustainability audits are done by consultants who offer sustainability auditing as a service for hire. Currently, most examine the carbon footprint of the organization, the operations and culture to see if sustainability ideas and policies are proactively managed, the procurement of supplies for operations and for product, and socioeconomic impacts. This type of internal, industry-focused sustainability audit identifies employment and income generated from building activities as socioeconomic impact.

Industry and the corporations and partnerships that manage them have been at the cutting edge of environmental regulation; however, they are not the only source of environmental impacts. The sheer volume of human population increases will generate large impacts on natural systems. The scope of environmentally regulated activities will expand under sustainability to areas of society other than big industry. It will spread to reach smaller industries, and then municipal emissions, military emissions and wastes, and other waste streams that impact the environment. Knowledge of the environmental impacts of development provides much of the impetus toward sustainability.

Audits

Like financial audits, environmental audits inventory and monitor a set of environmental and other factors to determine and improve performance, correct errors, and also to avoid liability.

Environmental audits may contain sensitive information that relates to some competitive advantage, or to a legal liability. Disclosure of such information sometimes discourages companies and organizations from collecting these data in the first place and thus forego a significant opportunity to plan and participate in improved performance. Some states and the federal government have created protection from disclosure of these kinds of information to encourage environmental and sustainability audits. These audits are entitled to a certain degree of protection as privileged documents. Privilege is a legal term protecting certain information from public disclosure because of some higher public purpose to be achieved by secrecy. Privilege and the extent of privilege are controversial.

Beyond environmental assessments emphasized by a NEPA process, some businesses and organizations are attempting to implement sustainability by auditing a wider set of consequences, and focusing attention on constructive ways to improve and restore ecosystems and communities in which they do business. NEPA and similar first-generation assessments and processes require mitigation of harms, rather than prevention and elimination. The shift toward prevention and elimination of harm rather than repair of damage already inflicted is one key principle in achieving sustainability.

A variety of creative tools and methods are now available to assist businesses that want to go further than mitigation strategies—toward sustainability. These tools include strategies to eliminate pollution and waste from production methods, eliminate toxins from products, reduce materials use and energy use in production, include and listen to a variety of stakeholders including community in decision-making processes, and strategic planning that includes true and full cost accounting for activities that impact future generations' ability to provide for their needs. ***See also* Volume 2, Chapter 4: Audit Privilege; Volume 2, Chapter 5: Sustainability Assessments and Audits.**

References

Bell, Simon, and Stephen Morse. 2008. *Sustainability Indicators: Measuring the Immeasurable?* London, UK: Earthscan.

Gibson Robert B. et al. 2005. *Sustainability Assessment: Criteria and Process*. London, UK: Earthscan.

President's Council on Sustainable Development (PCSD)

In 1993, President William (Bill) Clinton established an executive council to advise him about moving the U.S. toward sustainability along the lines of the Rio Declaration and Agenda 21. The PCSD became the chief voice formulating policy for sustainability in the United States. The formal power of the PCSD was limited to making recommendations to the president. Many of the members of the council, however, were heads of federal agencies and heads of prominent corporations and nonprofit organizations. As leaders of organizations, these members have some power to adopt and implement suggestions voluntarily, and many have tried to do so in the following years without governmental support or legislative mandates. For example, many corporations have followed Ray C. Anderson's path toward sustainability with his corporation, Interface. (See Ray C. Anderson and Interface Corporation, this volume, Chapter 2.) In addition, the U.S. Environmental Protection Agency under the leadership of Carol Browner embraced and maintained several programs aimed at assisting businesses in achieving better than compliance results, and moving toward several of the goals of sustainability including clean production methods, and transforming their industrial sectors through leadership.

The PCSD was asked to find ways to "to bring people together to meet the needs of the present without jeopardizing the future." Its final report issued in 1999 set forth a work plan for environmental, economic, and equity goals in the United States. This plan was shelved and largely reversed by the succeeding Bush administration. The PCSD left a rich blueprint for future action in a wide variety of areas in its last report, including recommendations for environmental management, urban and rural issues, transportation, clean production processes, development aid and its flow internationally, and clean energy development.

Programs for Business and Industry

Government provides numerous programs to assist business and industry in going beyond compliance with statutory minimums. Some of the most effective and creative are described next.

Design for the Environment

Design for the Environment is a government program bringing together businesses and industrial sectors and environmental groups for the purpose of reducing the use of toxic chemicals in products and production processes, and also achieving energy efficiency in these areas. This emphasis on pollution prevention and energy efficiency is an example of government exercising its power to convene stakeholders to seek voluntary change, rather than the traditional model of government as an investigator and enforcer of laws prohibiting certain amounts and types of pollution.

Reference

Bailey, Robert G. and L. Ropes. 2002. *Ecoregion-Based Design for Sustainability*. New York: Springer.

Sector Strategies

By focusing on particularly important industrial sectors, improvements in performance can be disseminated throughout a particular sector using the power of government to provide information and to convene stakeholders including state and local governments, businesses, and others. Some of their projects include:

- An attempt to quantify greenhouse gas emissions from the 14 key industrial sectors. This baseline of information is essential to approaching accuracy in predictions and risk management strategies.
- Promoting connections between businesses to enable waste from one business to be used in the production processes of another. This waste to food or fuel movement is a core principle of many of the key concepts of sustainability from industrial ecology and ecobusiness or ecoindustrial parks to the idea of biomimicry and a restorative economy. The sector strategy program has sponsored a local study identifying beneficial reuse of industrial wastes in the gulf coast region of the United States that is home to heavily

impactful industries in our economies including cement manufacture, chemical manufacture, oil and gas extraction and refining, construction and demolition, and iron and steel mills.

Small Businesses

Small and medium-size businesses make up 99 percent of the U.S. economy according to the U.S. Small Business Administration. Although many programs and ideas about sustainability focus on corporations and multinational big businesses, small businesses occupy a critical position in our economies for achieving sustainable business solutions. Small businesses may not have the money or staff to devote to developing sustainability measures for their business in the ways that a corporate giant like Wal-Mart or Nike can. In addition, small businesses may not have access to the capital to make significant infrastructure changes as publicly traded corporations do. Governmental programs to facilitate sustainability in these small and medium-size firms must address both the knowledge and technical gaps, as well as the financial gaps in order to mobilize changes.

One governmental program for small and medium-size businesses is aimed at energy efficiency. Energy Star for Small Businesses is a joint program of the U.S. EPA and Department of Energy that provides consulting and technical support to small businesses for achieving reduced energy costs.

References

EPA sectors strategy program www.epa.gov/sectors/.
Green suppliers network www.greensuppliers.gov/gsn/page.gsn?id=about.

STATE AND LOCAL INITIATIVES

Sustainability is often implemented on the local level; all environmental issues are in an important sense local. Agenda 21 calls for sustainability to be implemented at this level. What follows is a sampling of the flowering of local initiatives promoting sustainable businesses at the local level.

International Council on Local Environmental Initiatives

The International Council on Local Environmental Initiatives (ICLEI) provides policy advice to help local governments achieve their sustainability goals. It is the largest association of local governments in the world. Its membership includes more than 1,054 local governments representing over 300 million people in more than 68 countries worldwide.

ICLEI's flagship campaign, Cities for Climate Protection (CCP), boasts over 650 local government members worldwide who are beginning to make climate-change mitigation a part of their decision-making context.

CCP participant cities account for around 15 percent of greenhouse gas emissions caused by humans. Member cities have ICLEI guidance in a five-"milestone" process that lays out a flexible framework for cities to use. CCP participants are to first find a baseline for greenhouse gas emissions for one year and a forecast by which these cities can measure progress. Next, the city must decide what its emission-reduction target will be. Then, through a multistakeholder process, the CCP participant must develop a plan that describes exactly what the government will do to lower greenhouse gas emissions. The fourth step is to actually implement the plan. Finally, the CCP city will watch the progress to ensure that its goals are met.

Ekurhuleni, South Africa

Small measures can add up to create huge energy savings, as evidenced by a case study done on CCP participating city, Ekurhuleni. City leadership called for an energy retrofit of municipal buildings and premises. At the national level, energy efficiency is recognized as an inexpensive, but effective, means of meeting sustainable development demands. The municipality replaced 2,003 standard light bulbs with compact florescent light bulbs, added 23 mechanisms for instantly boiling water, lighting timers, LED light bulbs, and energy-efficient florescent lights. The total cost was just over $41,000; however, these measures produced a 53 percent, or $50,664, energy savings per year. The most impressive outcome is that CO2 emissions were lowered by 308 tons/year, sulfur dioxide emissions were lowered by 2,611 tons/year, nitrous oxide emissions were reduced by 1,226 tons/year, and total suspended particles were reduced by 100 Kgr/year. These results could occur easily in other cities.

Los Angeles, CA, United States

EnvironmentLA or Los Angeles's program for enforcement, climate change, sustainability, and green building is a comprehensive web connection to LA's Environmental Affairs Department, which is charged with carrying out Mayor Villaraigosa's sustainability initiatives (see www.lacity.org/ead/environmentla/index.htm).

LA's Million Trees LA is an initiative to increase the size of the city's canopy because its 21 percent coverage falls behind the national average of 27 percent. Furthermore, the plan clearly intends that the species of trees planted be native and drought-tolerant because of the hot climate. The city is committed to increase its fleet of alternative fuel vehicles by 15 percent per year. Under the EnvironmentLA, the LA Department of Water and Power has created the Refrigerator Turn-In and Recycle program where consumers can turn in their older model refrigerators and freezers that are not energy efficient for a rebate and compact fluorescent light bulbs. In turn, when residents purchase a new refrigerator, the city will give a rebate to those who choose to purchase an energy star efficient appliance. In 2007, LA issued a Climate Action Plan to reduce greenhouse gas emissions to 35 percent below 1990 levels by 2030. LA is approaching land-use issues through three main components of a green-building program: The Green Building Team, The Standard of Sustainability, and The Standard of Sustainable Excellence. In turn, these components address site location, water efficiency, energy and atmosphere, materials and resources, and indoor environmental quality. The Standard of Sustainability is a mandatory requirement that large projects, both new and certain existing ones, meet the intent of Leadership in Energy and Environmental Design (LEED) certified level. LA may be the model for how U.S. cities operate their business and economic development plans.

References

Gottlieb, Robert. 2007. *Reinventing Los Angeles: Nature and Community in the Global City.* Cambridge, MA: MIT Press.

Pavel, M. Paloma. 2009. *Breakthrough Communities: Sustainability and Justice in the Next American Metropolis.* Cambridge, MA: MIT Press.

Precaution and Local Government Initiatives

When harm to the environment or human health will result from development, reasonable measures should be taken to prevent harm even if the scientific evidence is inconclusive. This effectively shifts the burden of proving risk of harm from a member of the public or a potential victim of harm, to the proponent of activities that may cause harm.

This core principle from Agenda 21 has been applied by state and local governments to land use regulation. The National Association of County and City Health Officials adopted Resolution 03–02 on September 8, 2003, urging that the precautionary principle be applied to the activities of local health departments. This resolution illustrates the burden-shifting aspect of precaution.

> WHEREAS, the precautionary principle states, "When an activity raises threats of harm to human health or the environment, precautionary measures should be taken, even if some cause and effect relationships are not fully established scientifically. In this context, the proponent of an activity, rather than the public, should bear the burden of proof. The process of applying (this principle) must be open, informed and democratic and must include potentially affected parties. It must involve an examination of the full range of alternatives, including no action.

The resolution further suggests practical steps to implement the precautionary principle:

- Integrate public health and land use planning,
- Full participation by affected communities in land use decisions that affect them, and
- Better training of public health officials to increase their capacity to understand and participate in land use planning decisions.

Municipalities Adopting the Precautionary Principle

The precautionary principle is widely used in Europe and increasingly so in Canada. It directs governmental agencies to restrict the use of products or compounds suspected of producing health risks. The precautionary principle is articulated in Europe as a measure requiring industry to prove that a product or compound is safe rather than showing it is not harmful. This may just seem like semantics, but the articulation has the effect of creating requirements that are more stringent and protects the health of the public.

A municipality is a unit of local government in the United States. Towns, villages, cities, and other political subdivisions of the state are legally called municipalities. They have power over much of land use and land development. In the United States they have not been environmentally sensitive but tend to be focused on economic development through real estate. The state can preempt the land use of the municipality, and the federal government can preempt both the state and the municipality. Because the focus of municipal land use is generally economic development, it tends to make real estate transactions as efficient as possible to decrease transactional costs, such as those costs resulting from local regulations. Another factor is the highly held value of private property in the United States. This value also makes local land use regulation less involved in environmental issues.

Sustainability is directly concerned with the environmental impacts of past, present, and future local land use decisions. One of the ways that sustainability becomes involved with local land use decisions is through the application of the precautionary principle. The city and county of San Francisco (Calif.) were the first to do so in June 2002. Two municipalities that formally adopted the principle are the city of Berkeley, California, and Lyndhurst, New Jersey.

The San Francisco Experience: Sustainability in Local U.S. Government

Both the city of San Francisco and the county of San Francisco adopted the precautionary principle first in the United States. These are wealthy municipalities on the West Coast with highly educated and diverse populations. They are not especially industrial in terms of economic development but do have high concentrations of population. They are nestled between the Pacific Ocean to the West and mountain ranges to the East. The city of San Francisco is backed up into a large bay.

The adoption of the first local law, called an ordinance, was preceded by intense political dispute. It is likely to be copied by other municipalities, and the two discussed here did adopt significant parts of their language. Sections of this new law are directly related to developing a policy of sustainability, and often do more than adopt the precautionary principle alone.

The local elected body, called the board of supervisors, made specific findings to presage the law. These findings were that every San Franciscan has an equal right to a healthy and safe environment. This requires that the air, water, earth, and food be of a sufficiently high standard that individuals and communities can live healthy, fulfilling, and dignified lives. The duty to enhance, protect and preserve San Francisco's environment rests on the shoulders of government, residents, citizen groups, and businesses alike. These findings seem obvious to most people in most communities, but the local elected officials in San Francisco made them the basis for the adoption of the precautionary principle.

The San Francisco Board of Supervisors noted that historically, environmentally harmful activities have been stopped only after they have manifested extreme environmental degradation or exposed people to harm. It took special note of DDT, lead, and asbestos. In these cases regulatory action took place only after disaster had struck. The delay between first knowledge of harm and appropriate action to deal with it can be measured in human lives cut short. The board considered themselves a leader in making choices based on the least environmentally harmful alternatives. They specifically rejected traditional assumptions about risk assessment and management. San Francisco already had other local laws that dealt with specific policies and that included the precautionary principle. These laws included the Integrated Pest Management Ordinance, the Resource Efficient Building Ordinance, the Healthy Air Ordinance, the Resource Conservation Ordinance, and the Environmentally Preferable Purchasing Ordinance.

San Francisco elected officials sought to integrate all these environmental laws into one policy approach under the broad rubric of the precautionary principle. It also sought to have a more just application of these environmental laws. Its specific goal was to be sustainable by creating a sustainable San Francisco Bay area environment for present and future generations.

The city included a focused discussion of the potential solutions, problems, and limitations of science in their findings that presaged the precautionary principle ordinance. It found that science and technology are creating new solutions to prevent or mitigate environmental problems. It also found that science is also creating new problems with unintended consequences. In its view the precautionary principle is a policy to help promote environmentally healthy alternatives while preventing the negative and unintended consequences of new technologies. To do this, it uses a form of alternatives assessment based on environmental impacts. This assessment is based on the best available science. The alternatives assessment examines a broad range of options in order, with different effects of different options considering short-term versus long-term effects or costs. It then evaluates and compares the adverse or potentially adverse effects of each alternative, assessing options with the fewest potential hazards. This includes the option of doing nothing.

Another key principle of sustainability, transparent environmental transactions, is applied to the alternatives assessment under the precautionary principle ordinance. The alternatives assessment is a public process because, in the view of local elected leaders, the public bears the ecological and health consequences of environmental decisions. San Francisco's local leaders seek to fully engage the community in meaningful ways to the alternatives assessment process because they found that the final decision is more robust by broadly based community participation in alternatives assessment, as a full range of alternatives are considered based on input from diverse individuals and groups. The community should be able to determine the range of alternatives examined and suggest specific reasonable alternatives, as well as its short- and long-term benefits and drawbacks. This assumes the community has the capacity and time to do so. Citizens are assumed to be equal partners in decisions that affect their environment.

The local elected leaders of San Francisco based their findings for the new precautionary principle on the goals of a future where the city's power is generated from renewable sources, when all waste is recycled, when vehicles produce only potable water as emissions, when the San Francisco Bay is free from toxins, and when the oceans are free from pollutants. These goals are specifically founded in a hope for sustainability. They intend to use the precautionary principle as a means to help attain these goals and to evaluate future laws and policies in major areas as transportation, construction, land use, planning, water, energy, health care, recreation, purchasing, and public expenditure.

The San Francisco Board of Supervisors was also cognizant of the human behavior changes that will be necessary to live sustainably. It found that the precautionary principle as a policy approach will help San Francisco speed this process of human behavioral change by moving beyond finding solutions for environmental degradation to preventing environmental harm.

To implement the application of the precautionary principle, there must first be reasonable grounds for concern. This will be a matter of interpretation by city planners and managers. Where there are reasonable grounds for concern, the precautionary policy approach will be to reduce environmental harm by starting a process to select the least potential environmental threat.

This process is made of principles to guide the implementation and application of the precautionary principle, and these are some of the essential elements of sustainability. The first is anticipatory actions to prevent environmental or human harm. The stakeholders of government, business, community groups, and the general public share this responsibility and duty to engage in anticipatory actions, but these are as yet ill defined. If a particular stakeholder does know that an action will cause harm, then there is a duty to prevent it. This duty is based on knowledge, and another guiding principle of the precautionary principle law is an expansive right-to-know provision. The community has a right to know complete and accurate information on potential human health and environmental impacts associated with the selection of products, services, operations, or plans. Unlike many other right-to-know laws, the burden to supply this information lies with the proponent of a particular action, not with the community or government. There is also a strong duty to do full cost accounting in the assessment of alternatives to a proposed action. This is very extensive and includes raw materials, manufacturing, transportation, use, cleanup, eventual disposal, and health costs, even if such costs are not reflected in the initial price. Short- and long-term time thresholds should be considered when making decisions. Although not specifically stated in the precautionary principle ordinance, these could include the cumulative impacts of a particular alternative. At the end of three years, the local environmental agency, called here the Commission on the Environment, will evaluate the effectiveness of the precautionary principle ordinance and submit a report.

The San Francisco ordinance incorporating the precautionary principle as an overarching policy to local land, health, and environmental decisions is a new and foundational development in sustainable development. There are many questions about its costs, benefits, effectiveness, and interaction with other levels of government in the United States. Some view it as the U.S. catching up with European municipalities and their approaches. Others view it as economically unworkable, overburdensome to some stakeholders such as real estate development, and beyond the capacity of local government and citizens.

The Berkeley Precautionary Principle as Law

The municipality, or city, of Berkeley California specifically adopted the precautionary principle in its land use laws, called ordinances. Berkeley is a city near the city of San Francisco. Like San Francisco it has a highly educated, relatively wealthy, diverse population with some industry. A primary employer is the University of California at Berkeley, a leading U.S. research center. The specific purpose in the city of Berkeley in adopting the precautionary principle as an overarching policy is to protect the health, safety, and general welfare of the city. The law does this by decreasing health risks, improving the air quality, protecting the quality of the ground and surface water, decreasing resource consumption, and decreasing the city's impacts on global climate change. It intends to implement its precautionary principle law in phases.

The city defines its precautionary principle policy to mean where threats of serious or irreversible damage to people or nature exist, the lack of full scientific certainty about the case and effect shall not be viewed as a sufficient reason for the city to postpone measures to prevent the degradation of the environment or to protect human health. Furthermore, any gaps in the scientific data discovered by the examination of alternatives will serve as guides to future research on the issue. These scientific gaps will not prevent the city from taking actions protective of the environment. As new scientific information becomes available, the city can review these decisions and make any necessary changes.

The city of Berkeley intends to apply the precautionary principle under a set of guidelines. It will use anticipatory actions to prevent harm to the environment or to human health. It specifically states that government, business, community groups, and the public all share this responsibility to engage in anticipatory actions to prevent such harms. It is also guided by an extensive right-to-know requirement. The community has a right to know complete and accurate information on the real and potential health and environmental impacts associated with the selection of products, services, operations, or plans. This is one of the most extensive right-to-know laws in the United States. The city is also guided in its application of the precautionary principle by a requirement that all alternatives be fully assessed, and that the alternative with the least potential impact on health and the environment be selected. It must consider the impact of doing nothing as one of the alternatives. It is also guided by a full cost comparison. It needs to consider long- and short-term costs of any product, including an evaluation of the significant costs during the lifetime of the product. This includes raw materials, manufacturing and production, transportation, use, cleanup, acquisition, warranties, operation, maintenance, disposal costs, and long- and short-term environmental and health impacts. These costs are compared to any available alternatives.

The city's application of the precautionary principle is also guided by a participatory decision process that is transparent and uses the best available information. The city's current sustainable practices are incorporated into the new precautionary principle law. The city is to purchase products or services that reduce waste and toxics, prevent pollution, contain recycled content, save energy and water, follow green building practices, use sustainable landscape management techniques, conserve forests, and encourage agricultural biobased products. The city uses redwood that is certified as sustainably harvested. It will decide how the precautionary principle applies to future actions, based primarily on whether the city manager considers it feasible at the time. The new precautionary principle law also requires that an annual report on implementing actions be produced and available to the public.

The precautionary principle as urban policy in Berkeley is slightly more specific than the San Francisco law. It is reviewed annually, whereas the San Francisco precautionary principle law is reviewed every three years. Because the Berkeley policy is more specific and reviewed more, it may be a better research basis for analysts of sustainable development in U.S. local government. It is likely that state and federal levels of government will examine these laws closely. Some fear that state and federal government agencies will use their preemptive power to usurp these local initiatives before analyses of their effectiveness can take place. Others feel that the momentum of international sustainability movements will prevent the use of the preemptive power of higher levels of government.

Neither of these precautionary principle ordinances were formed in a particular controversy. They were both formed around a hopeful goal of sustainability for the future. Other older and more industrial cities also want to shape sustainable urban policies. These precautionary principles can be formed in the heat of environmental controversy.

Lyndhurst, New Jersey

Lyndhurst, New Jersey, is a municipality in the United States with heavy industry and environmental concerns about cancer clusters. After San Francisco and San Francisco County it was the second city to adopt the precautionary principle into law. In an effort to protect its citizens from the environmental impacts of heavy industry, the mayor and his staff developed a precautionary principle following international definitions and most of the San Francisco model.

The municipality of Lyndhurst, known as townships in New Jersey, adopted the precautionary principle as the policy of Lyndhurst. The city basically adopted the international statement of the precautionary principle, which is that when an activity raises threats of harm to human health, or the environment, precautionary measures should be taken even if some cause-and-effect relationships are not fully established scientifically. Although this is a broad principle, Lyndhurst narrows its application through its implementation. All its officers, employees, boards, commissions, departments, and agencies are required to implement the precautionary principle when conducting town business.

Lyndhurst further anticipates implementing the precautionary principle by developing new laws to create a healthier environment. It specifically wants to be a model of sustainable development by creating and maintaining a viable and healthy environment for both current and present generations. Its version of implementation of the precautionary principle is as both a policy tool and overall philosophy to advance environmentally healthy alternatives and remove negative and unintended consequences of modern and future technologies.

Another way Lyndhurst seeks to implement the precautionary principle is through an aspect of environmental justice. It wants to provide every resident with an equal right to a protected and safe environment and avoid disproportionate impacts. It will seek to find the least environmentally harmful alternative in every decision it makes. To use the precautionary principle for a healthy environment, Lyndhurst takes a strong ecological perspective of seeking a high standard of safety in the land, air, and water. It specifically wants to prevent environmental degradation and human health risks before they occur, not after they harm people or the environment. Whenever Lyndhurst is faced with a decision, it will analyze and assess all the alternatives. It intends to assess a wide range of alternatives and consider short- and long-range impacts.

Under this type of analysis, it may get to cumulative impact analyses. As Lyndhurst is an older industrial township, these impacts could be significant. The policy of the precautionary principle is to compare and contrast the adverse and potentially adverse effects of each option specifically noting alternatives with the fewest hazards. There are many environmental decisions with potentially adverse impacts that can include those not scientifically proven or disproved. Lyndhurst is sensitive to potentially hazardous activities and will ask if the activity is even necessary under the precautionary principle. It will try to measure how much harm it can avoid in its assessment of alternatives under their precautionary principle policy. This approach is unusual and a foundational development in U.S. sustainability. It places the burden of proving the safety of an activity directly on the proponent, and not on those opposing such activities such as the community or local government. If there is a threat of irreversible damage to the community or the environment, its policy explicitly does not accept the lack of full scientific certainty about the causes and effects as a legitimate reason to delay governmental intervention that protects the environment or the community. The township will revisit decisions as new scientific information becomes available. This may have the effect of increasing scientific knowledge and community-based monitoring of environmental decisions, both of which are necessary for sustainable development.

Like the Berkeley and San Francisco laws, the Lyndhurst precautionary policy has a set of guiding principles. First, there must be some reasonable basis for concern. Lyndhurst does have clusters of cancer and a history of heavy industry. Whether this alone is a reasonable basis for the application of the precautionary principle is unknown but would depend on the type of project or decision proposed. Nonetheless, if there is a reasonable basis for concern, then the precautionary policy creates a duty of anticipatory action to reduce harm. The law creates this duty in government, community, business, and the public. To anticipate environmental or community harm, it must first be known. The next guiding principle is a powerful right-to-know law that places the burden of creating the information on the project proponent. It specifically states that the community has the right to know complete and accurate information on the actual and potential health impacts of the alternatives under assessment.

And alternatives are required to be assessed, including the no-action alternative. An obligation is created to select the alternative with the least harmful impact. Alternatives are also to be evaluated under the principles of full cost accounting. This is a comprehensive cost accounting measure that evaluates all the costs including raw materials, manufacturing, transportation, cleanup, waste disposal, health impacts, and any others. It examines the entire lifecycle for the product or service, from cradle to grave. The entire process is to be transparent in that the community and public at large are to have meaningful involvement in finding and selecting alternatives under assessment.

Lyndhurst's precautionary principle policy is based expressly on the preservation of the environment. It creates a duty to enhance, protect, and preserve the environment on every citizen, business, nonprofit, and branch of township government. Under this law, the township wants to generate energy from renewable resources, recycle waste, and prevent pollution and toxins from entering its ecosystem. It views this law as providing the vehicle to develop future laws and policies that protect the sustainability of the ecosystem in the areas of land use, urban planning, water, energy, health care, recreation, transportation, government purchasing, and all public expenditures (such as education).

Lyndhurst developed its precautionary principle as law and policy in the context of a heated history of environmental and public health controversy. Court cases and scientists could not resolve this longstanding, simmering, adversarial standoff. Many municipalities are in this position in the United States and around the world. Economic development, capitalism, private property, and scientific postulations of cause and effect are now being directly challenged by communities and their democratically elected leaders. The result is a policy that goes beyond environmental preservation and conservation, beyond minimal protection of the public health, and that rejects current risk assessment models as inadequate. The result is a growing urban policy that is explicitly ecosystem-based, dynamically inclusive of community in meaningful ways, and that reorients local decision making toward precaution.

As these policies become implemented, they will create more controversy. Whether they are adopted by other municipalities, states or the federal government, or are contested in those arenas the controversy will continue. They are new, untested, costly, and may reveal unsavory truths about our past actions toward nature. ***See also*** **Volume 1, Chapter 3: Local and Regional Governmental Activities**

References

Myers, Nancy J., and Carolyn Raffensperger, eds. 2005. *Precautionary Tools for Reshaping Environmental Policy*. Cambridge, MA: MIT Press.

Whiteside, Kerry H. 2006. *Precautionary Politics: Principle and Practice in Confronting Environmental Risk*. Cambridge, MA: MIT Press.

Land Use and Growth Management

Land use refers to the array of government controls over the use of land, such as zoning. Zoning divides the land into allowable uses such as residential, commercial, or industrial. Land that government owns is also use controlled, such as parks, mines, and monuments. The U.S. federal government is a large landowner, and state and local governments can also own land. Private citizens and sometimes foreign corporations can also own land, known as private property. In the United States, if the land use control takes away all the value of a privately owned parcel of land, it is called a "taking" of private property, and the government must compensate the owner for the fair market value of the land. Many local and state governments try to regulate land use without it being a taking of private property. Takings of private property are expensive and often unpopular with citizens, land speculators, and real estate development.

Growth management refers to the timed and sequential control of land uses. Municipalities (towns, villages, cities, and counties) try to provide the necessary infrastructure for the type of land use growth they want. Infrastructure refers to bridges, roads, pipes, dams, and communication and power distribution. A community can control the kind of growth it wants with the infrastructure it chooses to provide, often called capital improvements projects. Developers often want the municipality to provide the infrastructure they want to maximize their profits. Sometimes they claim to have a right to force the city to provide the infrastructure. The question of who bears the cost of infrastructure improvements is complicated. Neither party wants the other to fail, but both want the other to pay for the type of infrastructure that meets their goals. Sustainable development is seldom the goal of developers, and occasionally the goal of cities.

Growth management of land uses has generally been devoid of serious environmental impact assessment until recently. Developers will mitigate environmental impacts only on site, if at all. These environmental mitigations are poorly monitored and therefore poorly enforced. The actions claimed as environmental mitigation are often short-term efforts, such as laying straw down on a construction site to prevent erosion of soil. There is no requirement to mitigate the impacts of environmental mitigation. So if the straw clogs the sewer and water system there is no requirement to mitigate that effect. Once the project is complete, the developer may be hard to locate, and the financing organizations for the developer may claim that the mitigation agreements are not enforceable. If, however, the project is financed with public funds, such as climate-neutral bonds, then some accountability may exist via the bondholders and the issuing municipality or state.

Growth management of land uses is fueled by population growth and the consumption of infrastructure by the current and new population. Population is increasing and so are potential environmental impacts. Alternative technologies in energy may decrease the rate of environmental impacts. One of the issues in the forefront of sustainable development is the control of land use growth to not exceed the environmental carrying capacity of the systems of nature on which future life depends. Some U.S. cities have recently implemented the precautionary principle in their laws and regulations, but it is not known if they apply to growth management of land uses. If one of the goals of growth management were sustainable development then aligning land uses with watershed practices could be a significant step for forward.

References

Barlett, Peggy F. ed. 2005. *Urban Place: Reconnecting with the Natural World.* Cambridge, MA: MIT Press.

Gottlieb, Robert. 2007. *Reinventing Los Angeles: Nature and Community in the Global City*. Cambridge, MA: MIT Press.

Helming, Katharina et al. 2008. *Sustainability Impact Assessment of Land Use Changes*. New York: Springer.

Kibbel, Paul Stanton, ed. 2007. *Rivertown: Rethinking Urban Rivers*. Cambridge, MA: MIT Press.

Climate-Neutral Municipal Bonding

Some states and local governments have begun to connect their considerable financial powers to the cause of mitigating climate change. State and local governments can raise substantial sums of money by issuing tax-exempt bonds. These tax-exempt bonds are attractive to institutional investors. Some municipalities and states now require that projects funded with these tax-exempt bonds must not increase greenhouse gases. This is termed as climate neutrality.

CHAPTER 4

Controversies

Controversies are part of any major change in ways of thinking about the environment. They are disputes between groups or individuals. In terms of sustainability, controversies are mainly about environmental issues and then about the policies necessary to mitigate impacts on natural systems. There are also major clusters of controversies around the business and economic impacts of sustainability and about the equity and fairness aspects of sustainable development. There will be many controversies surrounding sustainability.

As issues of environmental controversy become issues of sustainability, social issues will merge with environmental issues. Stakeholders are generally defined as those groups or individuals that have a stake in, or are affected by, the outcome of an environmental decision. Traditionally they were first environmentalists and industry. As government intervention grew to mitigate some of the environmental impacts of industrial expansion, the government became another stakeholder in many environmental decision-making processes. Under social policies of sustainability, even those environmentally based, stakeholders groups increase. State, local, and tribal governments are now developing environmental policies. Industries differentiate their stakeholders as big or small industries and as green industry versus resource extraction industry. Environmental groups are challenged directly and indirectly by environmental justice groups.

The range of environmental decisions that will be examined under emerging sustainability policies will expand into many past and present environmental decisions. This includes many decisions that were not considered environmental decisions in the past such as decisions that ignored cumulative environmental impacts, ecological impacts, environmental impacts on cities, land use decisions, and environmental impacts on public health. Risk assessment and monitoring are likely to expand. This may change the role of many social institutions such as higher education by requiring a greater emphasis on environmental assessment and monitoring in faculty and student development.

AGRIBUSINESS AND SUSTAINABILITY

Agriculture has been described as the systematic disruption of ecological systems to meet human needs. Agriculture's role in history was to produce food for a growing population. Agriculture's role under sustainable development is to treat natural systems sustainably. Sustainable agriculture is an important component of sustainable development. Some have questioned whether that can be done without degrading the environment.

Agribusiness, which practices agriculture and animal husbandry on an industrial scale, has provided food quantities that have established food security in developed nations and surpluses. Some types of animals adjust well to industrialized forms of animal husbandry and meat production; however, the effects of this scale of agriculture on farming families, their communities, and the webs of life have been high. Among these consequences are increased pollution of our land, air, and water, including water pollution, runoff, soil loss and salinization, vast increases of toxins in our water and food, and vast increases in animal waste. The increase in meat production alone may be responsible for as much as 22 percent of the human contribution to greenhouse gasses each year.

References

Bowler, I. et al. 2002. *The Sustainability of Rural Systems: Geographical Interpretations.* New York: Springer.

Clapp, Jennifer, and Doris Fuchs, eds. 2009. *Corporate Power in Global Agrifood Governance.* New York: Columbia University Press.

Kang, Manijit, S., ed. 2007. *Agricultural and Environmental Sustainability: Considerations for the Future.* Boca Raton, FL: CRC Press.

Quinn, Daniel. 1995. *Ishmael: A Novel.* New York: Bantam/Turner Book.

Robinson, Guy M. 2008. *Sustainable Rural Systems: Sustainable Agriculture and Rural Communities.* Aldershot, UK: Ashgate Publishing.

Agriculture and Soil

The soil is an integral component of agriculture. Human civilizations with good soil lasted longer than those without good soil. Good soil, as a natural system, is created very slowly but can be easily destroyed. Depending on climate, geology, and slope, soil is "created" at the rate of anywhere between .00067 and .00315 inches a year. This means it takes about 600 to 1,500 years to form 1 inch of soil. In areas where the topsoil is thin, agricultural practices can irreparably damage this natural system in a relatively short time of 500 years.

Reference

Montgomery, David R. 2007. *Dirt: The Erosion of Civilizations.* San Francisco: University of California Press.

Agricultural Technology and Soil

Along with civilization, technology increased the efficiency of agriculture. One of the primary instruments of agricultural technology is the plow. Most farmers in the world use it, mainly for sowing crops. The reason for plowing is to bury weeds with last season's crop remnants' and whatever soil amendments added in the course of the year. Plowing also opens up the soil, allowing exposure to air. This can accelerate some decomposition processes that release nutrients like nitrogen in the soil. Depending on climate and local conditions, plowing dark soil can increase the temperature in the soil. In northern and some southern latitudes, early warming of the soil can expand the growing season and range of plants under cultivation. Early soil warming can also accelerate germination times for some plants.

Plowing the soil, however, also allows for greater exposure to wind and rain. This can cause loss of valuable topsoils and erosion. Plowing is considered by many to be the main cause of degradation of agricultural soils. It also increases the vectors of ecosystem exposure from pesticides, fertilizers, soil sediment, and other wastes. As climate change expands desertification in highly populous areas around the equator, the loss of food production from agricultural land degradation in the context of still increasing populations pushes policies of agricultural sustainability.

Plows developed over history from sticks used to plant seeds, to the use of animals to pull stick-made plows, to iron-tipped plows. Later the mold board plow was developed. This plow had a curved blade that turned the soil over on itself instead of just tearing open a furrow to plant seeds. In 1837, John Deere invented the sod buster, a tough steel moldboard plow specifically designed to break up U.S. tall grass prairie soil. This soil was very dense with roots, and often gummy. When the tractor was introduced to agriculture, many plows could be pulled behind it, as well as other soil treatment technology like discs and manure spreaders. Treatment of the soil is called tillage, and it can be an expensive cost of doing business for farmers and create environmental impacts such as loss of topsoil. In the late 1940s and early 1950s, a series of herbicides were introduced to agriculture. These had the effect of reducing tillage, but some also had environmentally degrading and persistent impacts on the environment.

Plows and plowing the U.S. prairies extensively allowed for soil to be blown away in storms, causing crop failure and the Dust Bowl. It also had the effect of jump-starting U.S. land conservation practices. The Dust Bowl affected the United States from 1931 to 1939. This spurred the development of the U.S. Soil Conservation movement. The U.S. government formed the Soil Conservation Service, which is now the Natural Resources Conservation Service.

Natural Resources Conservation Service

The Natural Resources Conservation Service developed the Ecosystem Sustainability Framework for County Analysis. This framework is based on an agricultural perspective.

The Ecosystem Sustainability Framework for County Analysis project examines economic and environmental areas of agricultural counties. The model assumes an agricultural system is sustainable if it meets indicator threshold levels of sustainability. One observation that is part of this framework is that natural systems can be compared for individual resource concerns.

Reference

Waltner-Toews, David, James J. Kay, and Nina-Marie E. Lister. 2008. *The Ecosystem Approach: Complexity, Uncertainty, and Managing for Sustainability.* New York: Columbia University Press.

Sustainable Alternatives to Conventional Agriculture: No-Till Farming

Conventional plowing leaves the soil exposed to wind and rain, is prone to erosion, and spreads impacts to a large part of most ecosystems via agricultural runoff of pesticides and fertilizers, and sometimes topsoil. No-till farming promotes minimal soil disruption. In this approach, farmers leave crop residue on the field. They sow seeds with a specialized machine that makes a small slit for the seed. Leaving the soil undisturbed promotes soil stability and growth, reduces water needs, prevents runoff of pesticides and fertilizers, and decreases the costs of tillage in areas other than sowing seeds. In can also increase biodiversity for many species from worms to birds. A key point of no-till farming for sustainable development is that it can sequester carbon and decrease emissions of greenhouse gasses, thereby reducing the impact on global warming. Most crops take carbon dioxide out of the air and produce oxygen in a process of photosynthesis. The carbon dioxide remains in the roots and crop remnants. Between 1982 and 2003, according to the U.S. Department of Agriculture National Resources Inventory, soil erosion of U.S. cropland decreased 43 percent. Many claim this is a result of increasing adoption of no-till practices.

These practices, however, can lower crop yields and increase herbicide use. The increased moisture retention can introduce new diseases. In some instances, no-till farming may require more nitrogen at first. Without the soil-warming effects of plowing, no-till farming may also have slower and later germination at northern and southern latitudes. Some water crops, such as rice, are not yet appropriate for no-till approaches.

Agriculture, Soils, and Fuel

Should agriculture grow fuel? The use of corn-based fuels has greatly increased the amount of cropland devoted to corn. If biofuel crops are a wave of the future, soil sustainability will be a key factor in mitigating the environmental impacts from such a large increase in production demand. Corn, for example, is the basis of many currently new biofuels. When growing, corn does not get its nitrogen from the air, as many other crops do. As a result the nitrogen loading necessary for soils to sustainably raise corn may be larger.

Soil and Waste Capacity

Part of the ecological footprint of all human habitation is waste. Waste increases as population and patterns of conspicuous consumption increase. The type and amount of soil are also important to the amount and type of waste it can handle. The capacity of landfills is an important and often controversial issue when protecting soil-based natural systems from environmental impacts. As landfills reach capacity they can erode natural systems; however, waste must go somewhere. With the types and amounts of waste increasing, waste that is not recycled or reused is placed in waste transfer stations. These are temporary storage places, usually located along major waste transit routes. Typical waste transit routes are major roads, rail lines, and both inland and ocean shipping routes.

Because soil is a natural system, the impact of waste on it is a great concern for sustainability advocates. Soil is necessary for agriculture, but as landfills reach capacity, new landfill pressure may increase on agricultural lands. Two confounding issues are that agricultural industries can also produce large amounts of waste that degrade the environment, and sometimes "waste" from urban areas and the agricultural business can be reused either in the production process or in other industries or natural processes. Many cultures in the world use human waste as fertilizer, and animal waste as fuel and building materials. Organic wastes can be remediated by healthy soil. Chemical wastes can cause environmental degradation and may have to work their way through ecosystems.

Reference

Limitations on landfill capacity for the U.S. by state and region see soils.usda.gov/survey/geography/hurricane/index.html.

Irrigation and Drainage

Runoff nutrients contribute to water pollution and acidification. Insecticides, rodenticides, and fungicides add to crop yields but pollute water and can have harmful effects on farm workers, their families, and the ultimate consumer.

Farming is still among the most hazardous ways to earn a living with conditions largely outside the standards of protection given to other workers. Farm workers and farm families often have no health care provision outside of emergencies. Industrialized agriculture has also eliminated support services that provided the basis for local economies, leaving farm communities struggling to survive.

Recognizing some of these devastating environmental and community conditions, some governments provided subsidies to try to encourage healthy and environmentally suitable agricultural practices such as soil conservation, crop rotation, and family farming. Over time, many subsidies were given to support crops, such as tobacco, and unsuitable land practices such as irrigated crops in artificial settings, which can contribute to the salinization of soil. Recipients are often agribusinesses, not family farm operations. They often view subsidies as a form of entitlement and resist changes in policy regarding subsidies whether these changes are the result of free trade agreements or other policies.

Agricultural trends indicate that more farming operations will operate in a factory-type setting. This includes bringing more animals indoors for processing into food products. The scale of these factory-type operations has a much heavier ecological footprint on the place and the people who live and work there. Sustainable agriculture will not waste or pollute its ecosystems, will support its communities, and can maintain itself dynamically over time by adapting to changing circumstances.

References

Robinson, Guy M. 2008. *Sustainable Rural Systems: Sustainable Agriculture and Rural Communities.* Aldershot, UK: Ashgate Publishing.

Warner, Douglass Keith. 2007. *Agroecology in Action: Extending Alternative Agriculture through Social Networks.* Cambridge, MA: MIT Press.

Agribusiness

The business of producing and marketing food is often called agribusiness. Agribusiness has been remarkably successful at the production of enough food to ensure food security in developed nations. In these nations, public policies such as agricultural subsidies have encouraged both cultivation and conservation. Agribusiness has also been a huge consumer of chemicals such as pesticides, rodenticides, fungicides, and fertilizers. These chemicals have played an enormous role in ensuring the reliable cultivation of certain types of crops. They are applied to the land, but they end up in the water and the air, contributing to the environmental impact of farming operations in ways that detrimentally impact ecosystems. As it grew, agribusiness also overtook family-owned farming and smaller scale agricultural services that supported agricultural communities. Cities and urban areas have also expanded rapidly at the same time period, resulting in competition between agribusiness and urban areas for land and water. The challenge of sustainability for agribusiness will be how to integrate environment and sustainability into producing and marketing crops. The challenge for agricultural policy will be how to integrate environment and sustainability into food security policies and agricultural sector planning. These are the roots for several controversies about agribusiness. ***See also* Volume Two, Chapter 2: Key Business Sectors.**

Reference

Lyson, Thomas A., G. W. Stevenson, and Rick Welsh, eds. 2008. *Food and the Mid-Level Farm: Renewing an Agriculture in the Middle.* Cambridge, MA: MIT Press.

Subsidies

Public subsidies for the growth of certain types of crops have proven to be controversial in many countries. These agricultural subsidies violate free trade agreements like the World Trade Organization (WTO) and North American Free Trade Agreement (NAFTA). (See discussion, this volume, Chapter 3.) Efforts to negotiate an end to this practice, however, have failed so far. This failure is a setback for sustainability for different reasons. Sometimes these subsidies support unsustainable practices, such as the use of pesticides that harm both people and the environment. Other times these subsidies require the production of crops that are not usable by farmers for anything but trade. They can also require crops that use too much water for a given ecosystem and strain the carrying capacity beyond its natural limits. ***See also* Volume 2, Chapter 4: Resource Subsidies.**

References

Aksoy, Ataman M. and John C. Beghin. 2005. *Global Agricultural Trade and Developing Countries*. Washington, DC: The World Bank.

Meyers, Norman, and Jennifer Kent. 2001. *Perverse Subsidies: How Tax Dollars Harm the Environment and Economy*. Washington, DC: Island Press.

Monoculture

When agribusiness finds a crop that can be successfully grown and marketed, that product may well overtake the market in ways that squeeze out cultivation of other types of crops. When one type of plant takes over widespread cultivation, anything that particularly affects that cultivar will have a widespread effect on food security. Monoculturing of our food crops may make food supply ultimately more vulnerable. ***See also* Volume 2, Chapter 2: Biodiversity.**

Biodiversity: Monoculture and Food Security

Each ecosystem presents dynamic conditions for successful agriculture and for failure. Nature itself presents a broad range of plants uniquely adapted to these changing conditions. Humans have selected and manipulated the cultivation of these plants through agriculture. Agriculture cultivates plants for food, fiber, and fodder. Human selection of plants has narrowed based on market considerations including time to maturity, spoilage in transportation, labor handling, consumer preferences, and advertising. In addition, science now allows humans to genetically alter plants based on these preferences (genetically modified organisms). The power to select plants to cultivate has increased the amount of food available to human populations, but this choice has also eliminated the cultivation of many native plants.

Many native plants fit better into their present ecosystem because they can survive the diseases and climatic conditions present there. They may require less water and fewer pesticides, and may tolerate the presence of other plants. They can be grown closer to the point of consumption, requiring less transportation and decreasing environmental impacts. Native plants may also fill niches in the local ecosystem for life other than humans.

Nations that have monetary debt to organizations like the World Bank may have to grow nonnative plants for cash to pay these debts. These crops are known as "cash crops." As these nations grow dependent on money for modern items like cars, televisions, and refrigerators, their cash debt increases and their agricultural production focuses more on cash crops. Eventually they must import food and pay for it, which increases their debt and the production of cash crops that use nonnative plants. Many island nations rely heavily on imports for a modern lifestyle and have forsaken native crops for cash crops. The Western lifestyle in many contexts increases the environmental degradation of sensitive island ecosystems and vulnerability to famine by overreliance on nonnative cash crops.

Livestock Production

Under industrialization models, the production of animals for food has moved indoors, into factory-type facilities, replacing outdoor grazing and roaming. This indoor production of animals is called contained animal feeding operations or CAFOs. CAFOs produce enormous streams of waste that can pollute the freshwater ecosystems around them. In addition, the odor can be highly offensive at moderate distances. Finally, the treatment of the animals themselves may be deemed brutal.

The meat products generated by this type of factory production method may be treated with antibiotics to prevent the spread of disease in close, confined quarters. This use of antibiotics may be a contributing factor to growing antibiotic resistance. Hormones may be added to animal feed in these operations to maximize meat or milk production. Some of these hormones, such as bovine growth hormone, are controversial and have been challenged in the WTO process when other countries tried to ban products containing them.

References

Ansell, Christopher, and David Vogel. 2006. *What's the Beef? The Contested Governance of European Food Safety.* Cambridge, MA: MIT Press.

Choi, Euiso. 2007. *Piggery Waste Management: Towards a Sustainable Future.* London, UK: IWA Publishing.

Spellman, Frank R. 2007. *Environmental Management of Concentrated Animal Feeding Operations.* Boca Raton, FL: CRC Press.

Problems of Scale

The scale of agribusiness is often much larger than individual family-operated farms and ranches. When vast tracts of land are under irrigation or drained or cleared for large-scale agricultural use, the impact on local ecosystems is vastly different than the impact made by smaller farms. This can exceed the carrying capacity of the ecosystem and impact the lifecycles of various organisms in the ecosystem.

Reference

Hof, John, and Michael Bevers. 1998. *Spatial Optimization for Managed Ecosystems.* New York: Columbia University Press.

Urban Farming

Urban farming refers to farming, or growing, food in urban areas. It is generally difficult to farm in urban areas because of the lack of space, knowledge of farming, and security for the crops in the growing area. It is difficult to farm in urban areas in a way that makes a profit so that most urban farming is for sustenance.

Urban farming is making a comeback as traditional farming methods require transportation to get from field to market. Transportation costs add to the price of food. The lack of land is a challenge for urban farmers. There are also new farming technologies, such as vertical farming, that encourage urban farming. Some communities are beginning to provide land for community gardens. Security for the growing crops is important. People need the incentive to harvest their crops before they are stolen.

Sustainability is served by urban farming because it decreases the environmental impacts of food production, storage, transportation, and packaging. The necessities of farming, soil, water, and sunlight are often in shorter supply in urban areas. For this reason, cities pursuing sustainable development need to plan around their urban farming needs.

Organic Farming

Organic farming is a method of agricultural production that does not use chemical pesticides or fertilizers. Chemical farming methods expand the range of crops that can grow in a given area, expand the range of seasons to grow those crops, and increase the productivity per acre of land. As a result, they also help increase the profits of agricultural businesses and farmers. The increasing use of pesticides and fertilizers may negatively affect the environment when these products run off into the watershed. For example, one factor in the destruction of the Great Barrier Reef off Australia is thought to be the high nitrogen and phosphorus content of the water runoff from agricultural lands. They encourage algae blooms that can increase the water temperature and attract starfish that destroy living coral reefs. Chemical farming or conventional farming tends to use large-scale approaches in a way that is not site specific, ignores ecological balances, and poses threats to biodiversity. Organic farming is a better fit with the carrying capacity of the ecosystem.

Organic farming uses natural methods of growing crops, especially in the soil. It is generally more labor intensive. Part of the movement of organic farming was the idea of returning waste to produce soil that would in turn produce food. This cycle would, in theory, keep soil quality high even as population increased in the future. By preserving systems of nature on which future life depends, organic farming was tied to modern concepts of sustainability. One modern version of organic farming is biodynamic agriculture. In this system the soil and planets are life forces that work together to make the farm an integrated, holistic unit. Another modern version of organic farming is organic gardening. Organic farming approaches are applied to gardening approaches that raise plants for recreation, aesthetics, or landscaping.

In the United States, the National Organic Program enacted a definition of organic farming in 2002. The definition of organic farming is "A production system that is managed with the Organic Food Production Act . . . to respond to site specific conditions by integrating cultural, biological, and mechanical practices that foster cycling of resources, promote ecological balance, and conserve biodiversity" (Volume 7 of the Code of Federal Regulations, Part 205). Critics have charged that this definition increases the commercialization of organic farming and erodes its small farm, rural character.

Organic farming respects the systems on which future life depends and as such is part of sustainable development. However, it does not produce high profits, does require labor, and may limit agricultural productivity in some ecosystems.

References

Guthman, Julie. 2004. *Agrarian Dreams: The Paradox of Organic Farming in California.* San Francisco: University of California Press.

Organization for Economic Cooperation and Development. 2003. *Organic agriculture: Sustainability, Markets, and Policy.* Wallingford, Oxfordshire, UK: CABI Publishing.

Permaculture

Permaculture is a system for design within ecosystem capacities. Its early emphasis was on growing food within the capacity of a given ecosystem. It has since expanded to include building design, land use designs, and landscape restoration. The relationships of permaculture to ecosystems tie it directly to sustainable development.

Permaculture goes beyond organic farming because it looks at food in the context of the complete ecosystem and its natural range of biodiversity. By making the complete ecosystem more diverse and healthy, food productivity also increases. Food includes plants and animals. Birds are considered important parts of the ecosystem for their pest control, waste, seed propagation, and use as food. Chickens are often considered a part of permaculture if they are free ranging.

An important part of any ecosystem is the watershed. Most development uses impermeable surfaces that greatly increase runoff of water instead of letting it filter through the soil. The increased runoff can cause soil erosion as well as remove valuable soil nutrients. In nature, ecosystems absorb water through soil. Water in the soil also helps the soil become more productive by encouraging biological processes that can break down wastes into nutrients for plants. Ecosystems have different types of soil and different amounts of rain. Some ecosystems have soil that can retain more water than other parts where it may be rocky. Some ecosystems in dry areas may need to retain more water in order to grow plants and maintain soil-based biological processes. Permaculture will work with these carrying capacities of the ecosystem.

Permaculture is experiencing a rapid period of growth as sustainable development becomes a priority in many societies. For permaculture to be successful, an accurate and complete knowledge of the ecosystem is necessary, including human impacts.

Part of the recent growth of permaculture includes reviving environmental knowledge bases of traditional and indigenous societies, collecting and preserving older seed stocks and animals, developing urban planning principles, and building sustainable and intentional communities. Critics do not think that permaculture has proven itself to be productive enough to provide adequate food for a growing population. Some critics charge that permaculturalists bring in exotic and nonindigenous species into ecosystems to justify human food production.

Reference

Allen, Jenny, 2006. *Smart Permaculture Design.* London, UK: New Holland Publishers.

Vertical Farming

Vertical farming is a way to grow food in an increasingly urbanized world. Not only is the world becoming more urbanized, the population will increase dramatically as will the need for food. Food security will become a major issue. Traditional farming methods will not be able to keep pace. Some estimate that more than 80 percent of the land available for traditional farming methods is already in use. If organic and permacultural methods are used, there may be decreased productivity, further decreasing the ability of traditional agriculture to keep up with food needs.

Farming vertically allows for greater productivity, near the source of consumption, on a year round basis. It may also decrease the pressure on already environmentally degraded ecosystems and be flexible enough to handle climate changes caused by global warming. The amount of uncertainty in traditional agricultural food production because of climate changes is very great. This increases the amount of risk to food security.

There is some dispute about how much increased productivity vertical farming offers. It depends on the crop and management systems. Generally, one indoor acre is equal to about 4 to 30 acres outdoors. There is also less crop loss because of weather, and pests are more easily controlled. It may make organic farming approaches easier. Also, water is more efficiently contained. Some vertical farming advocates claim that it can create energy through the methane produced by composting organic matter. In terms of uses of nonrenewable energy, vertical farming produces food near the source of consumption, eliminating energy use in the distribution of food as well as energy use in the production of the food.

Vertical farming offers the potential to increase food productivity near population centers. Food security decrease is the first human sign that natural systems on which present and future life depend are eroding. Vertical farming will protect and preserve ecosystems to the extent that traditional farming methods become constrained by the use of vertical farming. The design of urban systems will have to change to accommodate vertical agricultural production. Sunlight, higher building stress levels, waste treatment, and waste transfer stations all have to be incorporated.

Reference

Bailey, Gilbert Ellis. 2008. *Vertical Farming.* Glacier National Park, MT: Kessinger Publishing.

AUDIT PRIVILEGE

Environmental audits as traditionally applied by industry are used to determine necessary environmental permits, finds energy savings, and reduce wastes. There is a controversy about whether they are privileged, or allowed to be kept confidential. Many states in the United States have laws that protect industry environmental audits. Some argue that if they were not confidential, industry would not do them. Others argue that all environmental impacts are important. From a sustainable development perspective, the traditional environmental audit is inadequate because the environmental information is not transparent, monitored, or placed in an ecosystem context.

Ideally, environmental audits assess the environmental condition and impacts of property, processes, and facilities in a systematic and objective manner. Traditionally, they are used as a management tool to follow environmental laws and regulations and correct or mitigate any problems. They can be used to find cost saving in energy use and in waste management. As used with sustainable development, environmental audits can be pushed to evaluate the potential for any activities that would irreparably harm the environment for future generations.

Environmental auditing began when the U.S. Environmental Protection Agency (EPA) began, in the early 1970s. It has evolved into a set of specialized management tools as applied to industry. Environmental audits are generally done internally, but they have become specialized and are now done by consultants. Currently, there is no legal rule that requires an external environmental audit, although a court decision could require it in a specific case. As the demand for sustainable development increases, more stakeholders from the community and from environmental groups want independent, external audits that go beyond minimal compliance with environmental laws. It is likely that environmental audits will evolve in this direction, such as the LEED (Leadership in Energy and Environmental Design) evaluations for sustainable building construction.

Industries have different reasons for engaging in environmental audits. They are done primarily to meet the requirements of environmental laws and rules and increasingly to examine their liability for cleaning up polluted sites. As pollution cleanup costs increase, industries must disclose them to potential investors as contingent liabilities, especially if they are to be publicly traded. Environmental audits can be specialized. The basic compliance environmental audit simply ensures that the industrial facility is following all the rules and laws. There are specific environmental audits of property transfers to examine the extent of liability for pollution and the requirement of due diligence to discover or reveal pollution in a real estate transfer. There is an environmental audit of internal operating systems that focuses on environmental risk management. Hazardous materials that must be followed from cradle to grave in their treatment, storage, and disposal have their own specific environmental audit. These environmental audits can be combined, and an overall audit is used by industry to determine what, if any, environmental liabilities need to be voluntarily disclosed.

Environmental audits can be mandated via law or court decisions, or voluntarily. Traditionally, they have been used in industry to avoid environmental controversies. Industry strongly prefers they be voluntary, but more communities are considering their own environmental audits of industry and of the operations of their own community. Environmental audits of complex organizations, such as some industries, universities, and government agencies, usually require trained professional people skilled in water, air, energy, and waste evaluations. Under sustainable development, there is a need to have independent auditors that pull together objective facts.

References

Ackerman, Frank, and Lisa Heinzerling. 2004. *Priceless: On Knowing the Price of Everything and the Value of Nothing.* New York: New Press

Widyawati, Diah. 2008. *Environmental Audit and Compliance: The Role of Audit Policies in Inducing Audit Adoption and Compliance Behavior.* Saarbrücken, Germany: VDM Verlag.

BUSINESS DUTIES TO FUTURE GENERATIONS

Core definitions of sustainability rest on the idea of duties to future generations. These duties include the duty to leave the webs of life intact so that future generations have the same opportunities to live and thrive as present generations have had. The precise identity of future generations cannot be known, so duties to them require abstract thinking at some level. This type of thinking requires present generations to imagine life beyond the probable limits of their own lifetimes. Human scale life expectancies do change, but the concept of a generation remains useful as an abstraction of long-term thinking.

Many of the structures of business accountability are tethered to very short-term periods. Publicly traded companies account for their profits on a quarterly basis. Their shares are traded in a public market that is reported on every moment that the market is open, on a global basis that is virtually every moment of almost every day of the year. Institutional investors and private individual investors may make critical decisions on these hourly, daily, and quarterly reports. In addition, most corporations conceptualize their main purpose as being the maximization of shareholder wealth, and their directors tend to put that duty before all others.

If the change toward sustainable processes and products causes near term losses, companies may be crippled financially and in terms of direction and motivation. Yet sustainability demands that we consider future generations in what we do today. Thinking long term is a serious challenge to contemporary business.

Some businesses that embrace the challenge of sustainability have turned to mission statements and strategic planning processes that help to explain long-term strategies and build trust and confidence in long-term plans. Some states are embracing this trend by allowing corporate charters and other legal documents like bylaws specifically to permit corporate decisions that benefit future generations.

Businesses that use resources or leave pollution in ways that diminish the health and well-being of future generations should consider whether they are conducting business in an ethical or fair way (see Volume 3). Climate change driven by human emissions will eliminate many plants and animals from our world forever. In developed countries, current and future children face life-shortening epidemics of obesity caused in part by food and product additives like transfats and some forms of plastic packaging, as well as unsafe exterior conditions to play in. These conditions are related in substantial part to the way we do business, the resources we use, the pollution we create, and the poisons we use and produce to make and distribute products.

Reaping the benefits of those methods of production now and leaving an inheritance of loss is an issue of intergenerational injustice. Sustainability unequivocally demands justice for future generations. Many businesses have not looked at their economic activities in this way. They may have recognized certain economic costs and risks to their current practices, but not to a future generation. That kind of limitation in accountability and vision leads businesses and whole economies to shift major, irreparable harms onto the backs of those future generations, as well as on contemporaries who they do not recognize, such as poor people and poor communities.

Reference

Dobson, Andrew, and Derek Bell, eds. 2005. *Environmental Citizenship.* Cambridge, MA: MIT Press.

The Ceres Principles

Ceres is an organization of institutional investors and environmental organizations that drafted a statement of principles intended to guide business toward sustainable environmental practices. Ceres organizations have used their collective financial influence to integrate the values of sustainability into their corporations. A similar effort is the Investor Network on Climate Risk, a project of the Ceres organization.

The following Ceres principles are a 10-point mission statement of environmental responsibility that each organization endorses:

- Protection of the biosphere
- Sustainable use of natural resources
- Reduction and disposal of wastes
- Energy conservation
- Risk reduction
- Safe products and services
- Environmental restoration
- Informing the public
- Management commitment
- Audits and reports

Today that mission statement has been condensed to this: "Integrating sustainability into capital markets for the health of the planet and its people." Member corporations commit to environmental self-audits and public reporting of environmental outcomes.

CAPITALISM AND SUSTAINABILITY

Capitalism is a term used to describe free market-based economies both at an ideal or theoretical level, and as a current operating global political economy. Sometimes the free market is also called the "private" sector. This implies a lack of control by government, although in actuality, it is the state that allows many businesses to exist. It is one of the most powerful and prevailing value structures in the world today, as well as in recent history. Many feel that it is poised for a period of robust growth, with far-reaching environmental consequences. In terms of sustainability, there are many fertile grounds for controversy because most versions of capitalism run counter to preservation of environment for future generations. As former Vice President and Nobel Prize winner Al Gore noted:

> Capitalism's recent triumph over communism should lead those of us who believe in it to do more than merely indulge in self congratulation. We should instead recognized that the victory of the West—precisely because it means the rest of the world is now more likely to adopt our system—imposes upon us a new and even deeper obligation to address the shortcomings of the capitalist economics as it is now practiced.
>
> The hard truth is that our economic system is partially blind. It 'sees' some things and not others. It carefully measures and keeps track of the value of those things most important to buyers and sellers. . . . But its intricate calculations often completely ignore

> the value of other things that are harder to buy and sell: fresh water, clean air, the beauty of the mountains, the rich diversity of life in the forest . . . the partial blindness of our current economic system is the single most powerful force behind what seem to be irrational decisions about the global environment.
>
> Al Gore, 1992, pp. 182–83

Capitalism as a basic concept usually incorporates several characteristics. Private producers of goods and services hire workers. They produce the good or service with the sole intent, and a legal responsibility, to make a profit. Any activity that reduces profits is suspect. Economic development becomes a strong value of the people and of the government. The free market development of capitalism becomes developed and supported by government. Corporations are creatures of the state in that they are allowed to form according to the laws passed by the government. The basic form of corporate structure allows corporations to invest money without individual liability for the debts they incur. This decreases the risk of many financial transactions and allows for entrepreneurs, or risk takers, to seek new markets and develop new technologies.

Capitalism as a complicated concept involves multinational corporations with budgets bigger than many nations. Sometimes these nations have natural resources, some that could be irreparably damaged and could be necessary for future generations. A large enough multinational corporation in search of profits might disregard the potential harms to a community or even a nation. For example, a large multinational corporation that mines gold in a very dry desert region might disregard the harm to the ecosystem, surrounding community, and even national economy if the profits from mining were sufficient. Gold mining takes water and discharges arsenic in the process. In a very poor country, capitalism exerted by multinationals may prevent fundamental principles of sustainability from being operationalized. The application of the precautionary principle, where the risks of irreversible damage to natural systems is first assessed, is expensive and time consuming.

Capitalism as a complicated concept includes materialistic values, a strong public policy emphasis on economic development, and higher rewards for the strongest competitors. Materialistic values are expressed when identity and status become functions of owning and purchasing. Consuming goods and services for the value of consuming, or social status, and without regard to environmental impacts is materialism. As one scholar on growth noted:

> capitalism itself has answered the demands that inspired 19th century socialism . . . But the attainment of these goals has only brought deeper sources of social unease—manipulation by marketers, obsessive materialism, environmental degradation, endemic alienation. . . . we

> have the freedom to consume instead of the freedom to find our place in the world.
>
> Clive Hamilton, 2004, pp. 112–13

National, state, and local governments around the world develop many public policies around economic development. While they are not all capitalistic, they all seek to improve the quality of life of at least present day humans. The role of government and the type of government intervention into the free or private market are greatly affected by values of capitalism. Governments can respond to strong political pressure from legitimate industry groups and carve out special exceptions or subsidies for industries that show promise of high profits under the name of economic development. Governments can also offer transactional security and privacy for business organizations such as multinational corporations, partnerships, subsidiaries, banks, and other lending institutions.

References

Gore, Al Gore. 1992. *Earth in the Balance: Ecology and the Human Spirit.* Boston: Houghton Mifflin.

Hamilton, Clive. 2004 *Growth Fetish.* London: Pluto Press.

Government Intervention in the Free Market Under Capitalism

Government intervention in most forms interferes with perceived profits. Any government regulation that provides for government contracts, grants, or subsidies, however, is sought after by industry. Governmental intervention in this manner is considered misleading by environmentalists because this type of government intervention distorts the true environmental impact of the good or service. Government subsidies, along with other avenues other than government power and influence, may exacerbate industrial practices that could jeopardize some of the basic principles of sustainability. Sustainability proponents argue that these government subsidies should go to industries that are trying or supporting sustainability.

At the local level, many towns and villages are seeking industry to improve employment as part of their economic development package. In a capitalistic approach to land and private property the local government will seek to streamline regulatory processes designed to protect the environment and public welfare, abate property and utility taxes for a set period of time, and clean up and give away industrial sites. When regulatory requirements are overlooked at the local level they can also be overlooked at the state and national level. There are many projects in the United States with significant impacts on the environment that do not file any type of environmental impact assessment. These regulatory requirements are often minimal and act to protect the environment. Without them, overdevelopment, pollution, and unsustainable economic development occur. There problem is that mispricing goods and services by ignoring environmental impacts of production and consumption causes a misuse of the natural resource base. This upsets sustainability proponents because part of sustainability principles is that systems of life on which we depend are kept in as good a state for future generations.

As capitalism and democracy gain world popularity, some aspects of capitalism may be reaching their environmental limitations. Some environmental amenities have no present value because there is no value to future lives. As countries, states, and cities grapple with growing populations wanting to consume more under capitalism and democracy develop public polices, they will also face environmental limitations. This includes the climate changes to be wrought by global warming, natural resource scarcity, and food shortages. Some current predictions have the greatest climate impact on the equator and surrounding tropical regions. Population and food supply rest on a precarious balance in many nations of this region. If drought and desertification continue along with population growth, then environmental limitations will be much more severe for future generations.

All these features work against sustainable development. Materialistic values in a capitalistic free market that does not place a transactional value on natural resources threaten to consume resources and damage ecosystems. In one way of stating it, the goose that lays the golden egg has been killed; that is, the environment that provides so many resources on which life depends was destroyed before knowledge of the damage was politically powerful enough to find expression in government. That is one reason why sustainability proponents advocate for the precautionary principle. The high demand on natural resources of a growing population in a new global economy burgeoning with capitalism and free markets has not been planned, contains many unknown dimensions, and causes many conflicts and controversies around values. Agenda 21 addressed sustainability in agriculture in the context of population growth estimates:

> By the year 2025, 83 percent of the expected global population of 8.5 billion will be living in developing countries. Yet the capacity of available resources and technologies to satisfy the demands of this growing population for food and other agricultural commodities remain uncertain. Agriculture has to meet this challenge, mainly by increasing production on land already in use and by avoiding further encroachment on land that is only marginally suitable for cultivation.
>
> Paragraph 14:1. United Nations Conference on Environment and Development, Agenda 21, U.N. Document A/CONF.151.26 (1992)

Government facilitation of capitalism as a value and a public policy undergirding economic development also works against sustainability because it supports unsustainable but politically powerful industrial development. This is particularly true in agribusiness. The small, self-reliant family farm is a romantic idea from the past in most industrialized countries. Subsidies of tobacco, cotton, milk, rice, and other crops ignore their environmental impact, but respond to political constituencies at the local, state, national, and sometimes international level. Other nations may choose to subsidize the crops of their own powerful political constituencies. All of these types of government that developed economic policies to attract and enable free markets to flourish may come at an environmental price.

References

Eckersley, Robyn. 2004. *The Green State: Rethinking Democracy and Sovereignty.* Cambridge, MA: MIT Press.

Hart, Stuart L. 2005. *Capitalism at the Crossroads: The Unlimited Business Opportunities in Solving the World's Most Difficult Problems.* Philadelphia: Wharton School Publishing.

Hawken, Paul, Amory Lovins, and L. Hunter Lovins. 1999. *Natural Capitalism: Creating the Next Industrial Revolution.* Boston: Little, Brown.

Heal, Geoffrey. 2000. *Valuing the Future: Economic Theory and Sustainability.* New York: Columbia University Press.

Kraft, Michael E., and Sheldon Kamieniecki. 2007. *Business and Environmental Policy: Corporate Interests in the American Political System.* Cambridge, MA: MIT Press.

Paragraph 14:1. United Nations Conference on Environment and Development, Agenda 21, U.N. Document A/CONF.151.26 (1992).

Global Capitalism and Its Relationship to Global Environmental Conditions

The planet has seen a vast expansion of economic activity, even more than population growth. Many environmental writers today believe this rapid and unchecked economic expansion is the main cause of environmental degradation. The world economy is better organized and now operates on a global scale. Environmentalists, sustainability advocates, and others are concerned that a large population on the verge of economic growth, primarily via some type of "capitalism," will increase the rate of environmental degradation to further decrease the potential for sustainability. As capitalism becomes less an economic theory but more a political cause, it causes a failure to recognize all the nonmarket costs of production. One of the main costs that are ignored under capitalism is environmental impacts. These can occur in the use of raw materials; in the industrial processes of production; in the shipment, storage, and use of the product; and in the disposal of the product. These ignored environmental impacts can accumulate and damage ecosystems. They can take the form of air pollution, toxic and hazardous waste sites, and polluted water. Environmental degradation and public health decline are also part of nonmarket costs. Capitalistic political economies, such as the United States, also support many production activities that produce serious and unaccounted environmental effects. Some examples of these are mining, logging, and ranching. Some industries are virtually environmentally unregulated and industry self-reports much of the information. Modern environmental advocacy groups and most approaches to environmental public policy are inadequate to the global challenges presented by current versions of capitalism. Capitalism practiced this way has socialized the harms and risks from speculation and exploitation and privatized the gains from these activities to corporations and their shareholders. Several recent citations to this proposition have to do with bailouts of banks and home mortgage lenders. This has led to the accusation that for some governments, "Banks are too big to fail, and homeowners are too small to bail." ***See also*** **Volume 2, Chapter 2: Globalization; Appendix B: The Equator Principles.**

References

Clapp, Jennifer, and Peter Dauvergne. 2005. *Paths to a Green World: The Political Economy of the Global Environment.* Cambridge, MA: MIT Press.

Costanza, Robert, Lisa J. Graumlich, and Will Steffen, eds. 2007. *Sustainability or Collapse? An Integrated History and Future of People on Earth.* Cambridge, MA: MIT Press.

Harnessing Economic Forces for Sustainability

Emerging versions of capitalism are developing new strategies for approaching the demands of society for a sustainable community. The free market is considered important for sustainability because of the innovation and technological advancement that is characteristic of emerging businesses and industry. There are several themes to these emerging theories. Most call for greatly increased efficiency from resource use. Increased efficiency can be in the form of reuse, recycle, and reduced use of natural resources, especially those that are nonrenewable. Industrial ecology focuses on the elimination of "wastes" in any production cycles, finding other uses for them and sometimes protecting profit. Many see an economy that moves from the production of goods to the production of services as the necessary step toward sustainability. Many service sector businesses, however, have a high hidden environmental and community impact. Service sector businesses may have high impacts in terms of their energy usage and carbon footprint. These are not as transparent as with a manufacturing company, but they are real. For example, a study of the environmental impact of conducting an Internet web search found that "a typical search generates about 7 g of CO2. Boiling a kettle generates about 15 g."

Also, these service sector businesses may not provide employment opportunities to communities without the necessary educational achievement levels to operate their businesses. When governments trade off employment opportunities for educational revenues from business, they benefit contemporary workers at the expense of future workers and employment opportunities. ***See also*** **Volume 2, Chapter 4: Market-Based Solutions to Environmental Degradation.**

References

Allen, Timothy F. H., Joseph A. Tainter, and Thomas W. Hoekstra. 2003. *Supply-Side Sustainability.* New York: Columbia University Press.

Bergh, J. C. van den, and M. W. Hofkes. 1998. *Theory and Implementation of Economic Models for Sustainable Development.* New York: Springer.

Esty, Daniel C., and Andrew S. Winston. 2006. *Green to Gold: How Smart Companies Use Environmental Strategies to Innovate, Create Value, and Build Competitive Advantage.* New Haven, CT: Yale University Press.

The Sunday Times January 11, 2009, technology.timesonline.co.uk/tol/news/tech_and_web/article5489134.ece.

Modern Capitalism

The global economy has grown roughly 5 percent a year, the U.S. economy about 3.5 percent. It is a fast rate of growth, primarily fueled by political capitalism and via multinational corporations. In 2019, the world economy could double in size at a 5 percent growth rate.

Why Is Growth Necessary under Capitalism?

Economic growth is a strong value that undergirds capitalism. The function of growth underscores high and fast growth of products. Economic growth is not limited to capitalism. Other political economies also value economic growth, as noted by one scholar:

> Communism aspired to become the universal creed of the twentieth century, but a more flexible and seductive religion succeeded where communism failed: the quest for economic growth. Capitalists, nationalists—indeed almost everyone, communists included—worshiped the same altar because economic growth disguised a multitude of sins. Indonesians and Japanese tolerated endless corruption as long as economic growth lasted. Russians and eastern Europeans put up with clumsy surveillance states. Americans and Brazilians accepted vast social inequalities. Social, moral, and ecological ills were sustained in the interest of economic growth; indeed, adherents to the faith proposed that only more growth could resolve such ills. Economic growth became the indispensable ideology of the state nearly everywhere."
>
> J. R. McNeill, 2000, pp. 334–36

Growth is traditionally measure by gross domestic product (GDP), although this measure has become more controversial as global concern about the environment has increased. ***See also* Volume 2, Chapter 2: Language of Sustainable Business.**

References

Elkins, Paul. 2005. *Economic Growth and Environmental Sustainability.* London, UK: Routledge.

Harris, Jonathan Mark. 2003. *Rethinking Sustainability: Power, Knowledge, and Institutions.* Ann Arbor: University of Michigan Press.

Hart, Stuart L. 2005. *Capitalism at the Crossroads: The Unlimited Business Opportunities in Solving the World's Most Difficult Problems.* Philadelphia: Wharton School Publishing.

McNeill, J. R. 2000. *Something New Under the Sun: An Environmental History of the Twentieth Century World.* New York: W. W. Norton.

The Economic and Environmental Consequences of Growth

What is the record of economic development, spurred by burgeoning capitalism and rapid globalization? For sustainability advocates of rapid and radical social changes, this is a key question. The answer for many in the world is cumulative environmental degradation and a widening division between rich and poor. Based on Gini statistics, the World Bank ranks countries with the greatest and least inequality between rich and poor. Among the 30 developed countries of the Organization for Economic Development (OECD), the United States and the United Kingdom saw the biggest division between the half with capital and the half without capital since records were kept.

The poor of industrialized nations usually bear the disproportionate environmental burden of this growth. It is difficult to persuade nations, communities, and peoples suffering a low quality of life to act in a sustainable manner for the sake of the planet. The record of ignored environmental impacts is just emerging, and most aspects of it are controversial. Multinational industries reliant on natural resources are exploiting the economic condition of nations suffering a low quality of life, especially in the areas of mining and petrochemical production. Environmentalists charge that they are moving fast because the world community is enacting tougher and more enforceable environmental restrictions on their activities. Industry responds that they are simply reacting to market demand. Environmentalists are now much more aware of global environmental actions. The strength of the environmental movement is its shared observations of nature, because in most places in the world, the only observers are the people who live there. The strength of industry, however, is its rapid growth and political power. More than half the largest economies in the world are multinational corporations, not nations.

Economic growth has increased the quality of life and negatively impacted the environment. As measured by GDP the world economy increased by a factor of 14 from 1890 to 1990. In the same time period, the global population increased by a factor of 4, water use increased by a factor of 9, sulfur dioxide emissions increased by a factor of 13, energy use increased by a factor of 16, carbon dioxide emissions increased by a factor of 17, and the marine fish catch increased by a factor of 35. In the first century of capitalism, economic growth increased rapidly and environmental impacts continued to be ignored. Of concern to modern-day sustainability advocates are the unknown cumulative, synergistic, and ecological impacts of such rampant growth because these types of impacts could irreversibly affect natural systems of land, air, and water on which all future life depends. If economic growth continues to increase, the concern is that even more environmental damage could occur.

One dynamic that many feel portends rapid economic growth is the mechanism of global financial markets, which often drive the financing for other markets. High profit growth is the primary value. One way this is observed is growth in market capitalization and the price paid for its stock. The expected rate of profit growth is very important. Losses often cause market value to decrease. In a competitive capitalistic market, most major financing industries need to finance highly profitable industries and corporations to ensure they meet their profit expectations. As finance markets are now global and operate 24 hours a day seven days a week in many places, the ability to quickly and irreparably exploit a weakness in a nation's environmental laws to overuse a natural resource in a unsustainable and profitable manner is much greater. The dynamic of finance markets lead many to believe that some nations could pursue a path of unsustainable economic growth as free markets grow into rapidly increasing populations. All cumulative environmental impacts to date, and what are now small environmental impacts, may increase in scale as population and capitalism increase. The threat of irreversible damage to systems of life like air and fresh water could increase dramatically.

References

De-Shalit, Avener. 1995. *Why Posterity Matters: Environmental Policies and Future Generations.* London, UK; Routledge.

Growing Unequal? Income Distribution and Poverty in OECD Countries, www.oecd.org.

Medard, Gabel, and Henry Bruner. 2003. *Global Inc.: An Atlas of the Multinational Corporation*. New York: New Press, pp. 2–3.

Sachs, Jeffery. 2006. *The End of Poverty: Economic Possibilities for Our Time.* New York: Penguin Press.

World Development Indicators. 2008. The World Bank, web.worldbank.org.

Recent Evidence of Economic Growth and Environmental Impact

There are many challenges to economic growth as capitalism of any type in many parts of the world. Other nations may have different values and measure gross national happiness. Many locations do not have addresses of personal or commercial property. Land ownership and recordation may be nonexistent in written form. Some value structures do not allow the forfeiture of a home for payment of a debt, perhaps because the property is communally owned, or landownership is unclear. In these types of nations, many lending institutions are hesitant to finance projects because without the security of clear ownership to property as collateral, they could lose profit. Other nations do not have convertible currency, or currency valuable or stable enough to be traded for other currencies on the world market. The relationship of the government to its people is also very different in many places. These three factors—lack of identifiable place, nonconvertible currency, and shifting and unknown nation/people relationships—pose as much of an obstacle for capitalism as they do for sustainability.

A serious challenge to economic growth, and therefore capitalism in many cases, is concern about the environmental impacts. Although the overall trends discussed in the preceding paragraphs describe events over a century, the question becomes more pointed when it addresses whether environmental regulations mitigated environmental impacts enough to be a basis for sustainability. Right now, most researchers answer no. By examining economic growth and environmental impacts since the rise of environmental regulatory regimes, there is still economic growth and environmental impacts, although they are lower than the growth of the world economy. Environmental impacts are still getting worse, but at a slower rate. From 1980 to 2005, according to the World Resources Institute and the Worldwatch Institute, the gross world product increased 46 percent. During the same time period, paper production increased 41 percent, the fish catch increased 41 percent, meat consumption increased 37 percent, energy use increased 23 percent, fossil fuel use increased 20 percent, the world population increased 18 percent, nitrogen oxide emissions increased 18 percent, freshwater use increased 16 percent, carbon dioxide emissions increased 16 percent, fertilizer use increased 10 percent, and sulfur dioxide emissions increased 9 percent. All these measures are fraught with some degree of controversy. They do show that with environmental regulation, it is possible to make a difference, but they also show that the difference is not sufficient.

A great concern to sustainability advocates is the characteristic of capitalism to fuel exponential growth. Each year's successive outputs and profits are invested to increase the rate and amount of production. So, too, does waste and environmental degradation exponentially increase. Unlike economic growth, which has no theoretical limit, however, the global ecology has definite limits to its capacity to regenerate itself. There is only so much pollution the climate can take before it becomes impossible to reverse the process. Once these capacities are reached, ecosystems may reach a tipping point of no return. Exponential growth can easily occur in a world of finance and leveraging of credit, but in nature exponential growth can overwhelm the capacity of ecosystems to regenerate themselves.

Market advocates claim that the market is self-correcting. They point to technology as capable of mitigating environmental impacts because it can reduce the amount of natural resources consumed and produce new products that affect the environment less. Right now, this is not happening fast enough for sustainability advocates. The latest research from five large West European and U.S. research centers examined these very issues. They concluded:

> Industrial economies are becoming more efficient in their use of materials, but waste generation continues to increase. . . . Even as a decoupling between economic growth and resource throughput occurred on a per capita and per unit GDP basis, overall resource use and waste flows into the environment continued to grow. We found no evidence of an absolute reduction in resource throughput. One half to three quarters of annual resource inputs to industrial economies are returned to the environment as wastes within a year.
>
> James Gustave Speth, 2008, p. 56

References

Smil, Vaclav. 2008. *Global Catastrophes and Trends: The Last 50 Years.* Cambridge, MA: MIT Press.

Speth, James Gustave. 2008. *The Bridge at the Edge of the World: Capitalism, the Environment, and Crossing from Crisis to Sustainability.* New Haven, CT: Yale University Press.

Svedin, Uno Britt, and Hägerhäll Aniansson. 2002. *Sustainability, Local Democracy, and the Future: The Swedish Model.* Emeryville, CA: Kluwer Academic Publishers.

Volk, Tyler. 2008. *CO2 Rising: The World's Greatest Environmental Challenge.* Cambridge, MA: MIT Press.

Zovanayi, Gabor. 1998. *Growth Management for a Sustainable Future: Ecological Sustainability as the New Growth Management for the 21st Century.* Westport, CT: Greenwood Publishing.

Classic Capitalism and the Environment: Implications for Sustainability

As political economies embrace both capitalism and democracy, many search for the original basis for capitalistic theories. Given their robust growth, sustainability advocates also examine these works for environmental or ecological content. Some of the foundational works of classical capitalism recognized the threat to the environment. One scholar noted in 1944:

> To allow the market mechanism to be the sole director of the fate of human beings and their natural environment . . . would result in the demolition of society. Nature would be reduced to its elements, neighborhoods destroyed and landscapes defiled, rivers polluted, military safety jeopardized, the power to produce food and raw materials destroyed."
>
> Karl Polanyi, 1944, p. 73

A fundamental critique of classical capitalism is the preference for the present over the future. The future is always unknown. The value of science to capitalism is in its ability to predict the future within a given probability and range of error. Sustainability expressly values future lives at least to the extent present lives are valued. Future generations cannot be part of current supply-and-demand trends or markets. Their number and value are unknown, controversial, and may not be quantifiable. When governments try to begin regulatory standards based on sustainability concerns for future generations, they often face this issue of the value of future lives. Because governments are often regulating corporations and their environmental behavior, and because corporations value profits over unknown future lives, corporations often resist any government regulation of their environmental behavior for purposes of sustainability. For example, one aspect of sustainability is the precautionary principle. Here the environmental impact of a proposed activity is assessed to determine if irreversible damage to a natural system could occur. If so, there is usually an exploration of what mitigative actions could and should be taken. All this environmental assessment takes time and ties up capital in what is perceived as a high-risk financial proposition. The emphasis of the present over the future under capitalism also resists expenditures of time and money that does not contribute to profit.

The question of a future value of a human life is a question that is answered in a different context right now. In legal cases of negligence, the value of a wrongful death is computed by juries. Alternatives to juries include legislative schedules or administrative compensation schemes.

Reference

Polanyi, Karl. 1944. *The Great Transformation*. Boston: Beacon Press.

CLIMATE CHANGE

The science around climate change is not controversial at the most fundamental level. Science has now established to a degree of certainty that human-driven climate change is occurring rapidly, with consequences that will radically alter the life systems of our planet. The kinds of change required to avert disaster are also not debated at the most fundamental level. We have to reverse the amount of greenhouse gases in our atmosphere, especially carbon dioxide from the burning of all fossil fuels.

A variety of measures and technologies are available to enable this change, but changes will distribute costs differently to different people. Some people accustomed to great privileges will need to reexamine their expectations. Others seeking more developed standards of living will need to lead change by developing alternatives to the current models.

References

Adger, W. Neil et al., eds. 2006. *Fairness in Adaptation to Climate Change.* Cambridge, MA: MIT Press.

Dimento, Joseph F. C., and Pamela Doughman. 2007. *Climate Change: What It Means for Us, Our Children, and Our Grandchildren.* Cambridge, MA: MIT Press.

Volk, Tyler. 2008. *CO2 Rising: The World's Greatest Environmental Challenge.* Cambridge, MA: MIT Press.

COST-BENEFIT ANALYSIS

Cost-benefit analysis is often used in all sorts of environmental decision making. It relies on the ability to assign accurate numerical values to the costs and benefits of any particular decision. This method appeals to many people as a means of avoiding subjectivity. It can be used to avoid discussions about moral values when values are in conflict or contested. Changes toward sustainability will often create tension when they require changes that impact those comfortable with current conditions.

Cost-Benefit Analysis Applied to Environmental Sustainability

Considering the relative costs and benefits of a decision is a way of trying to determine the value of that decision. This method of analysis often requires comparing the relative costs and benefits of alternatives and choosing the least costly alternative. In this mode of analysis, costs may be limited to financial costs alone or financial costs in a restricted timeframe. Similarly, benefits may be limited to financial benefits alone and restricted to particular individuals. The exclusions of broad definitions of costs and benefits may be made to try to simplify the calculation. It can be especially confounding to perform this kind of calculation when costs or benefits are intangible or incalculable for some other reason.

Another weakness in cost-benefit approaches to environmental decision making is the failure to include the health impacts on the public. These are difficult impacts to quantify, and it is even harder to prove causality. Nonetheless, the effects of an unhealthy environment do exist. In the case of air pollution and environmental regulation of ozone, one researcher noted the following:

> It is clear that the impacts of ozone exposure are grave. The body of evidence that ozone causes chronic, pathologic lung damage is overwhelming. At levels routinely encountered in most American cities, ozone burns through cell walls in lungs and airwalls, tissues redden and swell, cellular fluid seeps into the lungs, and over time their elasticity drops. Macrophage cells rush to the lungs defense, but they too are stunned by the ozone. Susceptibility to bacterial infections increases, possibly because ciliated cells that normally expel foreign particles and organisms have been killed and replaced

> by thicker, stiffer, non ciliated cells. Scars and lesions form in the airways. At ozone levels that prevail through much of the year in California and other warm weather cities, healthy, nonsmoking young men who exercise can't breathe normally. Breathing is rapid, shallow, and painful.
>
> Curtis Moore, 1988, pp. 187, 195–98

When this type of analysis is used, it may drastically undervalue ecosystems and their services precisely because these are unique and difficult to value. In the absence of any replacement for them, their value is limitless; however, infinite value might paralyze decision making. This kind of problem is endemic to ecological decision making, which affects many of the challenges to sustainability that we face.

Reference

Moore, Curtis. "The Impracticality and Immorality of Cost Benefit Analysis in Setting Health Related Standards." *Tulane Environmental Law Journal* 11: (1988):187, 195–98.

Historical Overview of Cost-Benefit Analysis

The first large-scale U.S. public policy to apply cost-benefit analysis was the Flood Control Act of 1937. Federal funds were distributed to flood risks based on whether the benefits granted by the federal government exceeded their costs. Cost-benefit analysis was applied to global issues in the 1950s when the World Bank used it to determine investments in developing countries. "Benefits" are the market value of goods and services received. Ecological benefits were not considered, but short-term benefits in terms of profits and single-generation, quality-of-life indicators that were measurable were counted. "Costs" were considered the value of goods and services that were foregone. Costs to future generations of loss of ecological services, such as biodiversity, were not considered.

Cost-benefit analysis as a foundation of environmental public policy is controversial when considered through the lens of sustainability. Most economists consider environmental policy as good public policy when it produces positive net present value. Ecological benefits are often hard to measure, and benefits to future generations seldom have present value. One large problem is how to discount future lives. It is impossible to know how many future lives of humans or any given species will exist. Therefore, some discounting of future lives is required by cost-benefit analysis of sustainable development. Issues of risky future outcomes are hard to handle under cost-benefit analysis when their probabilities are unknown and the values of future decision makers are unknown. Another problem is that it is hard to value something that is lost forever, such as the extinction of a species. Yet another problem for cost-benefit analysis is how to value something that has intrinsic value aside from human values. Many environmental economists do this based on a "willingness to pay" value. This value is dependent on knowing the tradeoffs between choices now and for sustainable development in the future. So much of the environment is just being discovered that values based on present environmental knowledge may be incomplete.

Reference

Moore, Curtis. "The Impracticality and Immorality of Cost Benefit Analysis in Setting Health Related Standards." *Tulane Environmental Law Journal* 11 (1988): 187, 195–98.

True and Full Cost Accounting and Cost-Benefit Analysis

True and full cost accounting raises challenges with regard to the accuracy of this methodology. In addition, reliance on this method may exclude or seriously undervalue intangible benefits because it does not grapple with the difficulty of assigning numerical or financial value to intangible things.

The webs of life in nature provide humans with many invaluable, irreplaceable benefits. When these benefits are not reflected in the price of goods and services sold, the market undervalues them. Goods and services will then be cheaper than they should be if these benefits were recognized and included in costs. Undervalued resources are subject to waste and ultimate depletion. Full value of these benefits would encourage sustainable behavior by discouraging waste and pollution.

Cost-benefit analysis is a method of calculating efficient human behavior. It is only as reliable as it is accurate about the true and full measures of costs and benefits calculated. Some costs are difficult to quantify such as environmental and ecological impacts, impacts on our webs of life, and intangibles. Exclusion of these costs and benefits, whether unintentional or intentional, makes the behavioral choices selected less efficient and less reliable. The greater the exclusions, the greater the margin of error.

Assigning Value to Intangibles: The Problem of Zero

When we do not know how to value something, we tend to frame its value as zero. Zero can mean vastly different things at the level of theory, from infinite value to valueless. Mathematically speaking, zero will be calculated as meaning valueless. This can lead to dramatically undervaluing costs and lead to major policy decisions reflecting the opposite choice from moral values.

Short-Term Accountability

Many of the drivers for businesses are based on the need to account for profits in short-term increments. Publicly traded corporations must account for their profits and losses on a quarterly basis, every three months. Decisions to invest in any sort of longer term may be unpopular and risk losing the financial support of shareholders. Some loss of support may be tolerable on a short-term basis, but the loss of larger investors and groups of investors is of critical importance and will often determine the path chosen by management. Large groups of investors have tried to instill ethical considerations into these capital management decisions in a variety of ways. Some have required the companies and organizations in which they invest to plan for sustainability in their basic assessments and decisions. Others have required their companies and organizations to account for their investment activities in an annual report format that considers the three dimensions of sustainability—economics, environment, and equity. (See Appendices in this volume, The Equator Principles, and the Ceres Principles.) For the most part, however, these investor requirements are still bound by a linear timeline that is exceedingly short—three months, to one year.

Longer term perspectives are commonly found. For example, gardeners often plant for a season, as well as a generation they will never see. Even businesses can formulate their sense of purpose in terms that transcend short-term profits, and even extend to future generations. The short-term vision often derives from the requirements of standards of financial instruments and accounting, but even these perspectives permit our views to be broadened over longer periods, for example in the logic of amortizations of decades.

References

Allen, Timothy F. H., Joseph A. Tainter, and Thomas W. Hoekstra. 2003. *Supply-Side Sustainability.* New York: Columbia University Press.

Heal, Geoffrey. 2000. *Valuing the Future. Economic Theory and Sustainability.* New York: Columbia University Press.

DEVELOPMENT LENDING: STRUCTURAL ADJUSTMENTS

Development projects are sponsored everywhere in the world from developed nations to developing countries. Lending to developing countries to promote development has special political, environmental, and social considerations.

Many developing countries were colonized by lending countries. Colonization exported human and resource wealth from the colonies and contributed to the wealth of developed countries in ways that were not based on market values or fair trade alone. Instead, the wealth transfer of colonial empires was based on markets underwritten by military strength and violence. The poverty of some developing countries and former colonies is attributable in part to this history of violent exploitation. Some development lenders to these countries have agreed at times to reduce or forego interest payments, partly in recognition of the hardships and injustice that loan payments may impose on struggling nations.

Human poverty and environmental degradation are clear risks to the stability of any country. Both can lead to enormous dislocation of population and destruction of lives and livelihoods, as well as property loss. Development is not possible under those conditions. Although risk assessments for financial ventures typically consider sources of political instability, as well as environmental instability, they do not necessarily adopt a proactive stance about the role of business in reducing or eliminating these sources of risk. Instead, loan conditions like structural adjustments have focused on budgeting and currency measures without regard to their environmental and social consequences. Private financial organizations have been slow to acknowledge a role for themselves in the elimination of poverty, and degradation of the environment, but some multilateral quasi-public institutions have embraced this role in their missions and practices. Three multilateral institutions—the International Monetary Fund (IMF), the International Development Association (IDA) of the World Bank, and the African Development Fund (AfDF)—have embraced the goal of canceling 100 percent of the debt obligation of the poorest heavily indebted nations. The Multilateral Debt Relief Initiative (MDRI), enacted by the major industrial countries through their organization, the Group of 8, will give debt relief to 34 countries, 28 of which are in Africa. More heavily indebted poor countries may also be included in the future.

The governments of the world, on the other hand, have acknowledged the poverty in their midst and committed to a work plan to address human poverty and engage measures to develop within the limits of our ecology.

Businesses focused on the need to show short-term profits often overlook a fundamental question: Why should businesses care about human poverty? Some businesses are moved as a matter of personal moral commitment. There are other important reasons for businesses to seek to eliminate human poverty, however, including the political instability created by poverty that can lead to failed states, expropriation of business property, breaches of contract, and political violence. These conditions make commerce dangerous and sometimes prohibitive in countries descending into a cycle of political chaos, disease, environmental degradation, and human poverty. ***See also* Volume 2, Chapter 2: Finance.**

Reference

Berne Union, Investment Data, www.berneunion.org.uk/bu-investment-data.html.

DISASTER RELIEF AND DISPLACEMENT

The problem of environmental refugees encompasses a number of other issues of displaced people: the effects of climate change and rising ocean levels, land use practices, population migration and settlement, jurisdictional and legal issues, and others.

An example is coastal flooding related to climate change. Poor and working class communities are often displaced by rising waters or tidal waves. In the wake of the destruction this leaves, communities are torn apart, and families must temporarily leave to find shelter and opportunities. This out-migration, however, offers a chance to revitalize the "cleaned up" area with private investment. The harsh result is displacement gentrification and the seizing of prime waterfront real estate from resident populations who can no longer afford to return to their former homes.

Gentrification displaces a lower income community from its existing neighborhood by virtue of an influx of higher income residents who are willing and able to pay more. This special type of gentrification is unique because the displacement occurs first and then the gentrification process takes hold. Natural disasters associated with climate change will create opportunities for disaster-related gentrification and displacement of environmental refugees nationally and internationally.

Africa and Disaster Relief

Poverty is a brutal context for any disaster. Natural disasters like floods, drought, fires, and earthquakes cause much more damage to human populations when poverty is widespread. When poverty is in the context of a history of colonialism and heavily indebted nations, as in sub-Saharan Africa, the damage to the ability of people to sustain themselves in way that allows future generations to exist is put in extreme jeopardy. Food security decreases rapidly and vulnerable populations suffer greatly, especially rural populations, women, and children. The typical aid response is to provide food, generally to refugee camps. In this way, the most dramatic impacts of any disaster are lessened. From a sustainable development point of view, however, other aid structures also need to be considered. The ability of an impoverished population to become self-sufficient enough to feed themselves and to engage in world trade may require more than food deliveries to refugee camps. Using non-governmental organizations and civil societies to distribute both food and the capacity to grow food is one way to restructure disaster aid. Meeting the millennium development goals in the context of any given disaster is one way disaster aid is restructured to meet the needs of the present without jeopardizing the needs of the future. These goals were developed without the assumption of a disaster, but were designed to be correctable in midstream.

References

Hannesson, Rognavaldur. 2001. *Investing for Sustainability: The Management of Mineral Wealth.* Norwell, MA: Kluwer Academic Publishers.

Paarlberg, Robert. 2009. *Starved for Science: How Biotechnology Is Being Kept out of Africa.* Cambridge, MA: Center for the Study of World Religions, Harvard University Press.

Sachs, Jeffery. 2006. *The End of Poverty: Economic Possibilities for Our Time.* New York: Penguin Press.

Smith, Keith. 2004. *Environmental Hazards: Assessing Risks and Reducing Disasters.* Andover, UK: Routledge.

United Nations. 2008. *Delivering on the Global Partnership for Achieving the Millennium Development Goals MDG Gap Task Force Report.* New York: United Nations. Online at www.un.org/esa/policy/mdggap.

ECOTOURISM

Tourism is often a significant source of income in developing countries. Tourists bring cash into economies that lack access to cash and can allow local people to benefit from the natural beauty and uniqueness of their location without destroying it for resource use. Tourism itself, however, has a significant environmental impact, from the greenhouse gases released by traveling to destinations, to freshwater resources tourists use in abundance for washing and other services, to the toll hikers and climbers who may take on sacred sites. Can ecotourism provide a sustainable source of development for communities located in environmentally fragile areas?

The International Ecotourism Society (TIES) defines ecotourism as "responsible travel to natural areas that conserves the environment and improves the well-being of local people." (TIES, 1990). TIES is the oldest and largest organization dedicated to ecotourism around the world. The organization specializes in supporting and advocating for ecotourism development in more than 90 countries and operates under five principles that minimize environmental, cultural, and ecological impacts:

- Build environmental and cultural awareness and respect.
- Provide positive experiences for both visitors and hosts.
- Provide direct financial benefits for conservation.
- Provide financial benefits and empowerment for local people.
- Raise sensitivity to host countries' political, environmental, and social climate.

Ecotourism involves nature-based travel that educates the traveler and is sustainable within the local environment and community. Nature-based travel involves the flora and fauna of the specific area and can mean human influenced as opposed to a completely wild area. Because visitation to natural surroundings occurs with almost all vacations, it is the education and local sustainability elements of ecotourism that distinguish it from traditional tourism. The greatest difference between ecotourism and traditional tourism development, however, is the focus on education, preservation, and conservation of native culture; ecosystems; and biodiversity.

Another important attribute of ecotourism is the involvement of the local community. Research shows that no matter how valuable the natural resources in an area that bring in tourists, if the local communities are not personally benefiting from the increased tourism, the natural resources will not be preserved. An example of this is in Africa, where game farm animals were being poached by local farmers and hunters. The locals were poaching the animals even though, by doing so, they were minimizing the interests of tourists to visit that area as a result of fewer game. The problem was that before the game park began, the local farmers had used the land for grazing and hunting to provide for their families. Once the game reserve eliminated their ability to graze and hunt for a living, the locals were benefited more from a dead animal sold illegally then leaving the animal alive. Once the state recognized the rights and needs of the local communities and recommended that game farms incorporate local businesses and communities through education, training opportunities, and by taking advantage of the resident's superior land and regional knowledge, however, the community started to benefit from the game farm being there. As a result, poaching diminished drastically. Therefore ecotourism also incorporates local communities and helps to develop these areas in a culturally sensitive way. Ecotourism incorporates the environmental conservation, cultural sensitivity, and economic incentives to build sustainable businesses.

Although there is little uniformity among all the locations as to practices, there have been numerous international events centered around ecotourism development, standards, and goal. In 1992, with the Rio Declaration on Environment and Development, "the Rio Declaration", ecotourism got its foothold as a sustainable means of developing a country's tourism market. Many of the principles and goals of ecotourism are incorporated into the Agenda 21 initiatives. Also, in 2000, the Convention on Biological Diversity Guidelines on Biodiversity and Tourism Development was implemented to help states better understand how and why to protect ecological and animal diversity. The most important recognition so far of how important ecotourism is for the future of the planet, however, occurred when the United Nations named 2002 as the International Year of Eco-tourism.

Since 2002, there has been a burst of international ecotourism initiatives, local programs, and certification programs. Some of these include the Environmental Code of Conduct for Tourism, Green Labeling for Tourism, Responsible Travel Guidelines for Africa, Certificate in Sustainable Tourism for Costa Rica, and most recently, The Ecotourism and Sustainability Conference 2008 addressing greening the tourism industry in the United States and Canada along with many others.

Ecotourism has created controversy around issues of sustainability. The controversies start with issues of the impact ecotourism has on the environment, on indigenous peoples, and on the local population. There are concerns that ecotourism may decrease the biodiversity of sensitive environments. An increased exposure to people may alone affect the way plants and animals interact in their environment. The increase in waste that often accompanies more tourists may change the behavior of some animals that find it easier to live from wastes than to hunt food. The propensity of tourists to take souvenirs may decrease the actual biodiversity of an area. In some cases, the tourists and the guides may not be fully informed of the environment they are touring and the unintended consequences that may result on a given ecosystem. This is the case in some sensitive tropical rainforests where not all species are yet known. There is some concern that ecotourism may expedite extinctions by direct impacts, and indirectly by introducing invasive species or diseases that eradicate native species.

The relationship between indigenous people and their government may be tenuous. Many governments in areas populated by indigenous people seek economic development through ecotourism. One example of some of the dynamics of ecotourism is in Malawi, a country in Africa. The largest lake there is Lake Malawi. Fishing from these freshwater areas is the main food source for this very poor country. The biodiversity of the fish population is very high. On Lake Malawi is the fishing village of Mdulumanja. In this town is a large tourist hotel. The land, once owned by the country, was sold to a private corporation in 1987. The new owner insisted on tearing the village down to expand the hotel. He then stopped allowing the residents to use his water sources and also stopped buying fish from them. He put up a big fence all around his property. In 1990, after much dispute, he evicted the residents with the help of the national government. More than 70 long-term residents were forced to leave. The hotel owner razed all their homes for tourist development. He expanded his hotel to accommodate the tourists who were mainly white South Africans. The residents lived on the fish from that lake and now had to find other means without access to the lakefront. Ecotourism exploitation is dynamic. Lake Malawi National Park is a world heritage site and may be under the same threat of ecotourism development as the hotel industry marches along the waterfront. There is no monitoring of the impact of this ecotourist industry on the aquatic biodiversity.

The type of job creation under ecotourism is generally in the service sector. Maids, cooks, maintenance workers, drivers, and guides who serve the tourists follow the initial jobs in construction. These are generally low-wage jobs but in places of poverty are considered to be better than nothing. In some places they may exploit cultural and social oppression. The people trained and hired for these jobs may not be from the area. Tourists may not be aware of local cultural dynamics and can unintentionally further political and social oppression. In nations where the indigenous peoples have separate interests, ecotourism may occur as economic development that disrespects native cultures.

Many of the construction practices around ecotourism may not be sustainable, and may have negative environmental impacts on natural systems on which present and future life depends. In places where indigenous peoples have interacted sustainably with their environments in holistic ways for long periods, important environmental knowledge could be lost.

References

Bramwell, Bill. 2004. *Coastal Mass Tourism: Diversification and Sustainable Development in Southern Europe.* Someset, UK: Channel View Publications.

Hepworth, Adrian. 2008. *Wild Costa Rica: The Wildlife and Landscapes of Costa Rica.* Cambridge, MA: MIT Press.

Mowforth, Martin, and Ian Munt. 1998. *Tourism and Sustainability: New Tourism in the Third World.* London, UK: Routledge.

United Nations Environmental Program. 2008. *Climate Change and Tourism: Responding to Global Challenges.* New York: United Nations Publications.

ENERGY

Energy is what transforms natural resources into products for use in our households and businesses. Energy is how our systems of communication and information function. Energy is how basic functions within the built environments happen, from heating, cooling, and lighting structures, to processing materials for manufacture. The energy source that fuels our households and businesses is overwhelmingly derived from burning fossil fuels. Fossil fuels are nonrenewable resources. Not only do we know that they are finite and cannot be replaced when they are exhausted, we also know that they are responsible for terrible damage to the ecosystems that support all life. The climate changes that human activities using fossil fuels have caused are clearly signaling harm and danger to living beings and systems. The challenge for human economic activity is to stop using this source of energy. Transitional and alternative energy sources need to be developed in ways that make them useful.

The problem is not the lack of alternatives. Instead the problem is the cost of change, including changes of infrastructure in the developed and developing worlds. Ironically, developing nations may be in the better position to adapt and implement new technologies. Developed economies reliant on fossil fuels are faced with the difficulties of trying to adapt existing infrastructure to new technologies. Adapting existing systems is not a particularly effective way to change a system.

In addition, producers from agribusiness to neighborhood printers and individuals in their households have made substantial investments in the technology that they rely on for their daily activities. Those investments are expensive and are often calculated to pay off after a lengthy period. The need to change that technology is an expense that may lie beyond the ability of individuals or businesses to pay. Resistance to change for individuals, households, and businesses comes partly from the fact that obsolescence of so much of our technology and infrastructure at the same time seems financially unsupportable. The scale and cost of change, together with the urgency of need, can easily inspire a state of paralysis rather than mobilization.

Some entrepreneurs and economists see the challenges in different, more manageable terms. For example, the Natural Step and its proponents describe the business case for change as one of maintaining a profitable edge through changes toward sustainable production and products while resources are undergoing compression.

Sustainable development requires businesses and industries to adopt clean manufacturing goals and technologies. This often means designing pollution and waste out of their manufacturing cycle (industrial ecology) and thinking about their product in terms of its total lifespan, beyond its point of sale (product lifecycle management).

Clean production may also require a substantial investment in new technology and plants, investment that is prohibitive to small business enterprises. These requirements challenge businesses and industries in virtually all sectors of an economy to change what they are doing and how they are doing it. What makes the task of change even harder is the fact that many established businesses and industries have been subsidized directly or indirectly through tax benefits conferred by national governments, and these benefits enhance the reluctance of businesses to change.

Alternative energy sources to fuel our economy include ideas for using grains to make alcohol-based fuels. The most famous of these is ethanol made from corn. Corn is currently used primarily as a food crop for people and animals. Switching its use to an energy source has one effect of increasing the price of food, dramatically in some countries. The impact of increased food prices is felt most painfully among the poorest members of these countries. One suggestion is to use revenues raised from carbon taxes to subsidize the price of food.

Many alternative forms of energy are controversial according to local circumstances. Wind power uses large turbines to generate electricity, which is then transmitted to a grid of wires connecting providers and users of electrical power. The location of these turbines can be a source of local controversy. Some locations are desirable because of the strength and reliability of prevailing winds in the area; however, those locations may be sacred to some cultures, or simply considered inappropriate for development for aesthetic reasons.

Similar controversies surround the development plans for some sites for geothermal plants, hydroelectric plants, and solar electric plants.

References

Collin, Robert. W. 2008. *Battleground: Environment.* Westport, CT: Greenwood Press.

Knechtel, John. 2008. *Fuel.* Cambridge: MA: MIT Press.

Mez, Lutz. 2007. *Green Power Markets: Support Schemes, Case Studies, and Perspectives.* Essex, UK: Multi-Science Publishing.

Cape Wind and the Kennedys

Cape Wind is a wind generation project to be sited in the Nantucket Sound. The project would construct 130 wind turbines off the coast of Cape Cod. It is estimated that this would be a source of clean energy for 75 percent of the area, replacing coal-fired power plants that generate an estimated million tons of greenhouse gas emissions each year. The problem is that the wind farm is projected to be located in a part of the sound that is a favorite sailing and yachting location of the Kennedys, and it would rest within five miles of their home. Although the Kennedys have been environmental advocates, and advocates for change for generations, they have used their considerable political power to oppose this project.

ENERGY SOURCES

In terms of sustainability, energy is divided into two categories—renewable energy sources and nonrenewable energy sources. Coal, oil, natural gas, and uranium are nonrenewable energy sources because they will run out and are therefore considered nonsustainable. Also, their by-products are carbon dioxide and radioactive waste, which are not readily reabsorbed back into natural systems without damage to the environment. As our supplies of these nonrenewable energy sources run low and population and energy demand increase, retrieving and transporting these energy sources increase cost, international conflict, and environmental impact. For example, drilling for oil in the Arctic National Wildlife Preserve becomes more controversial as the international price of oil increases.

Many controversies exist around these traditional energy sources. Some claim that new technology will make radioactive waste nonhazardous. Others claim that oil can be retrieved and recycled from current uses. Besides energy sources, nonrenewable energy sources such as oil also form the basis for many products that whole societies are based on. Others claim that carbon dioxide can be captured and sequestered, decreasing its effects on global warming and climate change.

Renewable energy sources are considered sustainable. These include the sun, wind, and tides. These sources do not have waste products that harm the environment. Other energy sources are between renewable and nonrenewable. Alternative fuels such as ethanol from corn, biomass converters, geothermal heat wells, and hydropower from dams are sources that may be more sustainable than traditional nonrenewable energy sources. Some dispute this claim because these sources often require more energy than they produce and can produce waste products or activities that have negative environmental impacts.

Aspects of energy conservation are currently major parts of some sustainability programs, such as green buildings. Decreasing energy use does decrease environmental impacts, but if the energy source is not renewable then with population growth it may not be sustainable. This is the subject of some controversy. With green buildings and sustainability certification programs, energy conservation can mean using natural lights, water-permeable paved surfaces, tree shading, and provisions for alternative transportation.

Unsustainable Energy Sources

Contemporary economies are dependent on petrochemical and fossil fuels: oil and coal. These energy sources have significant costs for the environment including spills, dumping, drilling, and mining. Industries that rely on technologies built around these sources of energy will need to make substantial investments in technology and personnel if they are going to change. This scale of change is hard to imagine in the absence of a widespread disaster or war.

Industries built around these fossil fuels employ vast numbers of people. If these industries change, those workers will need jobs in other areas. The transition from unsustainable energy sources to more sustainable ones may cause serious hardship among people and communities employed or reliant on oil and coal.

Among the many alternative energy sources that could replace oil and coal, some choices have controversial impacts as well, such as nuclear energy. The waste associated with the creation of nuclear fuel poses serious risks and has not been resolved in a way that is ecologically sustainable. Nuclear waste is still not a closed ecological loop with the waste providing the basis for future processes. In addition, radiation poses hazards to workers and to communities reliant on this form of energy.

Sustainable Energy Sources

Many alternatives sources of energy are available. Most are generated by local conditions, rather than importation of a commodity. A few offer sufficient power to be exported to other locations. Their generation and use for energy may raise local issues of concern. They all operate without the creation of waste and pollution, based on current knowledge.

Solar Energy

Sunlight can be turned into electricity by solar energy technologies. When sunlight hits a solar panel the energy in the sunlight frees the electrons in the solar cells. This produces an electric current. In the usual crystalline silicon cell the electrons are tied tightly to atoms. When sunlight hits them the electrons they move more readily, creating electricity. The current power efficiency of crystalline silicon cells is such that about 20 to 25 percent of the sun's energy is converted to usable electricity. Currently mass-produced solar cells have only 15 percent efficiency. The highest efficiency solar cells are on satellites and are almost at the 50 percent range of efficiency. Here on Earth the highest efficiency of converting sunlight into electric power is 31.25 percent, set in 2008 at Sandia National Laboratories and breaking the old record of 29.25 percent set in 1984. There is much less interference from the atmosphere and more direct sunlight at higher levels in the atmosphere.

Access to sunlight is restricted by natural cycles of light, darkness, and local weather conditions. This source of energy is free, although the technology to convert it to human uses on a scale useful to modern appliances and fixtures can be a significant cost to business and individuals. Right now the costs of mass production are prohibitive because the raw materials and manufacturing processes are costly. New development in low energy devices (LEDs) such as lights can be applied with solar energy in poor areas of the world to provide light in otherwise unlit human habitations. There is much research in this area, especially in the use of plastics and of nanoparticles.

Urban landscape design, building codes, and environmental land use planning can help facilitate the access and use of solar energy. If buildings and other structures block the sun, then access to direct solar energy is denied. Buildings need to be built to hold solar panels and to gain the most from sun exposure. Environmental land use planning can help maximize access to sunlight by requiring alternative energy use, as occurs in parts of Germany.

In terms of environmental sustainability, solar power holds great potential. It is likely that further technological advances in solar power technology will increase its efficiency and decrease the energy impact on natural systems. Solar power also has the potential to reduce the environmental impacts of the energy distribution system because it can be produced near the point of use.

Reference

Bradford, Travis. 2008. *Solar Revolution: The Economic Transformation of the Global Energy Industry.* Cambridge, MA: MIT Press.

Hydroelectric Energy

When water flows and falls, it creates tremendous energy. The force of these flows and falls has the potential to provide energy for many human uses. Technology to convert this energy to human uses can begin by stopping the flow or fall and rechanneling its energy. Humans and other animals have used a variety of ways to do this by building dams from all kinds of materials. Beavers use mud and debris from their ponds. Humans use cement and metal constructions to capture the most powerful rivers and falls.

When waters have been sequestered by dams, the particulate matter that they transport will drop to the floor, forming layers of gathered materials called sediments. These sediments can concentrate particulate matter from runoff including highly toxic materials. When dams are removed, one consequence is to wash this concentrated sediment downstream to human settlements and to the oceans or seas.

Because human settlements tend to form around water flows for the purpose of transportation, blocking these flows often has the known consequence of inundating preexisting human settlements.

Falling and rushing water can generate substantial energy that can be harvested and used by humans. Water must be diverted and constrained through a technological device that uses or stores its energy. Ancient waterwheels, as well as contemporary dams, rely on the same technological idea of using water to power wheels to power machinery. The diversion of water and its constraint may create significant local consequences, depending on the scale of such interferences. Dams may result in inundation of whole geographical features as well as cities and towns. Loss of these areas and the human cultures they contain in order to generate power is controversial.

Fresh water is an increasingly scarce resource and an irreplaceable natural system in a world with an increasing need for it by an increasingly consumptive human population. Its use as an energy source when combined with gravity may or may not detract from other uses such as irrigation. Local or small power production from falling water is another way to be more self-reliant and may have less of an impact on the environment than a traditional energy source relying on nonrenewable rouses such as coal.

Geothermal Energy

Below the crust of the Earth, the Earth itself is in considerable dynamic thermal flux. Sometimes this thermal condition erupts at the surface in the form of volcanoes and geysers. Less dramatically, the Earth's hotspots manifest locally as steam and superheated conditions that can also generate energy sources useful for human activities.

From the perspective of environmental impacts, current usage of geothermal energy seems to have minimal environmental impacts. For the most part, people are simply gathering the heat that would leave the Earth anyway. One issue is how local or regional the energy is. Can heat be converted to energy such as electricity and be used over a long distance? Another issue is how large a human impact a geothermal energy source could sustain and how large a population and ecosystem it could support.

Wind Energy

Winds are powerfully driven across the land and water by their dynamic interaction. The power of these winds can be destructive and overwhelming in the form of hurricanes and tornadoes. Some places and conditions of wind, however, are highly predictable and offer the possibility of beneficial human use. Ancient uses of wind power include wind turbines.

Winds often travel in predictable directions at known places. This makes them good locations for windmills. One of the locations currently being explored for a turbine-type of wind power is a large bridge in the city of Portland, Oregon. The current bridge is old and needs to be rebuilt. The wind from the Columbia River Gorge is strong and predictable. An emerging concept that may increase the predictability of wind is the use of lasers on the wind turbines. These lasers would be aimed at the prevailing wind direction and reflect off particles in the air. When winds are less predictable, power production can ebb, and sometimes gusts can destroy the wind turbine blades. Currently, windmills are also explored for use at a residential level. They have been used in rural agricultural areas to draw water for wells. Like solar power, windmills need access to wind and require building and zoning code flexibility. Wind energy may also be more efficiently captured with future technology in windmill design.

Tides Energy

The sun and moon exert powerful forces on the Earth in the form of the rising and falling levels of oceans and some larger bodies of water. Over time, and in certain locations, the force of these changing levels can be harnessed to produce energy for human uses.

Tides rise and fall around the globe twice a day. In places where tides are constricted by land masses, they can rise and fall more than 40 feet. This is a large amount of water and energy. There are times when tides are still so that tidal power plants are operational only about 10 hours a day. This means that in comparison to other longer exposure energy sources, tidal power may be considered inefficient. So far, only about 20 tidal energy sites in the world have been identified, but some researchers are confident that there are more.

The concept of tidal power generation works like traditional hydroelectric power generation. Basically, a dam is built across a river estuary. In tidal energy terms this dam is called a barrage. It is generally much wider than a traditional dam across a river. The barrage has tunnels built into it that move turbines that create electricity. The turbines are turned by the tide coming in and then going out. The only tidal power plant in Europe is in France in the Rance estuary. It is the largest in the world and was built in 1966.

There are some environmental concerns with tidal power generation. They are built and operate in sensitive aquatic ecosystems. They decrease the normal rate of flow of the water and could upset delicate ecosystems there. Fish, amphibians, and birds could be influenced by the flow alterations.

Other ideas about the creation of energy from tides include the application of windmill-like structures as underwater turbines to create electrical energy. They would require high underwater current velocities or engage continuous ocean currents. Unlike traditional windmills, these turbine blades could operate in both directions to capture energy from ingoing and outflowing tides. This idea is still in the experimental phase.

Tides continue to attract researchers interested in developed alternative and sustainable energy sources. They are reliable and contain large amounts of energy. Their impact on systems of nature on which future life depends is not known to be threatening. The impact of moving large amounts of electrical energy through coastline estuaries is not known. Tidal energy may not be viable in places where tides are small, such as the equator.

Biofuels for Energy Cogeneration

As materials from plants and animals decompose, they release gasses and heat that can be used as fuel by humans. Capturing and converting these decomposing materials as gas and heat in a form that is useful to humankind are sometimes referred to as cogeneration of energy.

Biofuels generally refer to ethanol, butanol, and biodiesel. They replace petrochemicals as an energy source and come from plants such as corn. There is concern that the growth of plants necessary for the creation of ethanol will have negative environmental impacts. About 3 percent of our current gasoline supply is ethanol. Corn, for example, requires more nitrogen fertilizer than other crops and runoff of this fertilizer can overwhelm lakes and rivers. It is estimated that it takes three gallons of water to produce one gallon of ethanol from corn versus 2.5 gallons of water to produce one gallon of gasoline. If other plants are used to produce ethanol some claim the water usage to create one gallon of ethanol can decrease to one gallon of water.

The Energy Policy Act of 2005 requires fuel producers to nearly double sales of ethanol-blend fuel from 2006 to 2012. Currently, ethanol is mostly used as an additive to gasoline in low blends up to 10 percent ethanol and 90 percent gasoline; however, these blends may actually increase air pollution. They can increase ground-level ozone pollution by increasing emissions of nitrogen oxides and volatile organic compounds that help create ozone. These low ethanol blend fuels also decrease the emission of carbon monoxide, which slightly contributes to ozone depletion.

The development of biofuels as a sustainable energy source is dependent on the types of plants used to create the fuels and how the fuels are used. There are many types of plants that can be used to create biofuels, such as wheat grasses and kudzu. Plants could be genetically created specifically to make fuel. For sustainability purposes the question becomes one of the environmental impacts of these plants in the creation, growth, harvesting, and transformation into fuel. Do these plants and processes threaten systems of nature on which future life depends? Other questions of sustainability depend on how the fuels are used and how they impact their environments. Do they simply

Biomass as Renewable Energy Source

Concern about climate change and increased interest in sustainability initiated a search for renewable energy sources that are alternatives to petrochemical sources. The use of biomass for power, heat, and fuels is receiving the endorsement of nations like Germany for use as an alternative energy source. Biomass can be composed of many items, some grown especially for energy use: biomass forest timber, fast-growing tree species, and sometimes specially grown cereal straw. Biomass can also be the traditional waste products of other industries. This especially interests sustainability advocates who believe in natural capitalism because it closes the loop by making a by-product into a usable product. Biomass can include waste from agriculture, logging, untreated waste wood and sawdust, and some waste from food production. Wood is the primary biomass fuel. As a fuel, wood is considered carbon neutral because it can only release the carbon it received as it grew. Wood is used for site-specific heat generation, as well as for heat and power in medium-size industrial plants in Germany.

Biomass in the form of split logs, wood chips, and wood pellets feed ovens and boilers. These systems use electronically regulated combustion systems. Some of these systems use an auger screw to feed pellets for burning. These systems produce much lower emissions than stoves and fireplaces. In many places, reliance on woodstoves for heat pollutes the air with particulate matter as a result of inefficient burning. Larger biomass heating systems used to supply several homes are usually fed with wood chips from machine chipped wood. In larger facilities, biomass in the form of wood is used both to create heat and to generate electricity. The simultaneous generation of heat and electricity increases the overall efficiency of the biomass consumed. The modern wood pellet boilers automatically clean themselves. Ash disposal and servicing are needed about once a year.

Reference

Roser, Dominik et al. 2008. *Sustainable Use of Forest Biomass for Energy.* New York: Springer.

replace petrochemicals and emit the same pollutants that cause global warming and climate change? These are difficult, controversial questions. Many amateur and professional scientists are working on these issues.

New Sources of Energy

It is possible that energy sources not thought possible could exist and be the source of sustainability, or at least unlimited energy. The energy needs of the world's cities are expected to increase even more than the increases in population. Professional and amateur scientists are exploring ways to create new energy sources in their laboratories and even in home garages.

Algae Energy

The development of ethanol from corn started investigations into the use of other sources of biofuels. Algae require water, sunlight, and carbon dioxide to grow. Some scientists are experimenting with ways to make algae grow faster, or bigger. There are many types of algae, over 100,000 strains. Each strain has a different amount of carbohydrates,

Beetles and Ethanol

The search for sources of energy now incorporates biomimicry, copying processes of nature. A basic problem of getting ethanol from wood is the removal of cellulose from the wood. Cellulose is required to begin the biofuels production process. Wood contains a chemical called lignin, a polymer that is designed by nature to protect trees from insects. Human processes of removing lignin from wood require acids or expensive boiling processes. Asian long-horned beetles, however, have a larval stage in their lifecycle in which microorganisms in their digestive tracts can separate the lignin from the cellulose. Scientists are now studying the microorganisms to determine how this occurs. If it can be copied effectively, then another source of energy may be available.

proteins, and oil. As algae grow, they produce oil that can be easily converted into biodiesel and ethanol by fermentation of the algae's carbohydrates. Algae grow very quickly if the right conditions are met and can be harvested much more frequently than corn. Promoters of algae say it can produce more than 10,000 gallons of oil per day, much more than any current plant sources.

Algae production of biodiesel and ethanol may not necessarily decrease environmental impacts on the air if rates of fuel usage increase. To produce algae requires use of brackish water and some require large amounts of plastic. There is also some concern that if a type of algae is genetically created and escapes into the natural environment, it could cause ecological damage. Research is at an early stage, but many home-based entrepreneurs are forming companies to develop this product, pursue patents, and develop new markets.

Nuclear Fusion

Of all the energy sources known today, nuclear fusion is one that would reduce carbon emissions to zero and thereby decrease global warming and the rate of climate change. Today, nuclear energy is produced through nuclear fission, a process by which atoms are split. Nuclear fission releases energy when atoms are split, as nuclear energy is produced today. Nuclear fusion occurs when atoms become fused, releasing a huge amount of heat energy. The sun is powered by a sea of nuclear fusion reactions. Nuclear fusion has been replicated here on Earth with the highest temperatures in the known solar system. It cannot be sustained very long, however, because it is so hot. It does create nuclear waste but unlike the nuclear waste from nuclear fission, which cans last 20,000 years, nuclear fusion waste has a half-life of between 10 and 12 years.

Currently it takes more energy to create it than it produces but international research is underway to develop ways to control the heat and sustain the fusion reaction. This project is called ITER (International Toxicity Estimates for Risk Assessment). It is a large international project under construction to study the plasmas in conditions that will produce electrical energy. The members of ITER are the European Union, the Peoples Republic of China, the Republic of Korea, the United States, Japan, India, and the Russian Federation. It will be constructed in a town in southern France called Cadarache. Their energy goal is to produce 500 megawatts of fusion power for extended periods of time. If current computations are correct, this is 10 times the energy than is necessary to keep the fusion reaction going.

There are grave concerns about nuclear fusion. Some fear that it represents overreliance on a technological fix. Others fear that it could set off unknown catastrophic consequences on Earth. Still others think the scale and cost are so great that not enough energy can be produced quickly enough for a burgeoning world population.

Animal and Human Waste

Humans, like all other animals, produce waste. Waste has an undeniable effect on ecosystems and, when concentrated and untreated, it may pose a threat to systems of nature on which future life depends. As the human population increases and becomes more urbanized, waste treatment becomes a higher priority. Left untreated and ignored by social stigma, human waste can accumulate in parts of nature. Waste treatment and sanitation are hallmarks of a civilized society. Higher levels of human sanitation have contributed 20 years to the human lifespan. Before then, human waste was a vector to many diseases that killed the vulnerable old, young, and weak. Currently, 40 percent of the world's population does not have access to a place to excrete their waste, and 90 percent of the human waste in the world ends up untreated in lakes, rivers, and oceans.

The potential to create energy from waste has been explored and is applied to cattle. Cows produce methane that can be reused as energy. Methane gas is a by-product of animal and organic waste. It is a large part of natural gas. Natural gas is used to power sewage treatment plants and run other power generation plants. To use human and animal waste, known as biosolids, the moisture and carbon dioxide must first be removed.

Energy from animal waste often comes from cows. They were first used in U.S. dairy farms in the 1970s when oil prices increased. When oil prices and federal support for alternative energy sources declined, the effort stopped. Now that oil prices are again on the rise, energy deregulation allows the production of electricity from nonutility sources. State and sometimes federal government grants provide assistance to farmers who are again using animal manure to create energy. Traditionally, animal wastes have been moved to a lagoon or holding pond until spring, and then possibly used as a type of fertilizer on the fields. In this process they emit many greenhouse gases. Most of these farms creating energy use some type of "waste digester" mechanism and process.

Many of these are homemade and based on the ingenuity of the farmer. The energy content of animal manure alone is low so it is occasionally mixed with other higher energy wastes. Water washes the waste down into an influent tank where anaerobic bacteria degrade the biomaterial. Cow manure is about 65 percent methane. It takes about two to –three weeks after which the result is either sent for storage or used in a natural gas engine. This energy is often used close to the point of use, so that the distributional problems of energy creation are lessened. It is estimated that the 1 billion tons of cow manure produced per year in the United States cold generate 88 billion kilowatt per hours of electricity. This is about 2.4 percent of current U.S. annual consumption. It also decreases greenhouse gases by about 99 million metric tons. The material left over is more easily used as fertilizer. Farms also produce food, and to do so while creating energy and reducing waste impacts on the environment is seen as a hopeful development for many advocates of sustainability.

Human waste for energy creation has been used in China. San Antonio, Texas also plans to use human waste for energy. San Antonio is a rapidly growing city in Texas with a shrinking water aquifer. As water aquifers shrink, they tend to decrease in water quality. San Antonio is seeking to protect its remaining water quality and to harvest the methane from human waste for energy rather than risk the waste further contaminating the scarce fresh water resources. Currently, San Antonio produces 140,000 tons of human waste a year that will be converted into 1.5 million cubic feet/day of natural gas to run its power plants.

Energy from human and animal methane produced in this manner still faces the problems of environmental impacts of energy distribution to point of use. Other methods of waste to methane production, however, can be smaller scale with "bio mass" home converter systems. These biomass "digesters" convert human and animal waste and organic wastes into methane. They use bacteria to break down the wastes and convert it into methane gas that can be used or stored for energy in the unit.

Social attitudes about privacy, defecation, and other aspects of waste do not prevent the impact of waste on the environment. These systems may have other environmental impacts, such as carbon dioxide emissions; however, it is an energy production method that can be both centralized and localized, and reduce the flow of untreated wastes into water.

Reference

Rogers, Heather. 2006. *Gone Tomorrow: The Hidden Life of Garbage.* New York: New Press.

Direct Human Production of Energy

Although not yet practiced on a large scale, some places are beginning to experiment with the direct production of energy from people power.

For decades, many parents fantasized about having their children earn the power necessary to watch television by capturing the energy from their exercise. This idea is moving from fantasy to reality as new entrepreneurs begin production and distribution of exercise machines that do just that. Team Dynamo Bikes creates four connected bicycles that generate power. If they are used consistently, the company estimates they will produce about 1.5 kilowatt hours per day. The energy produced by the bicycle riders charges a group of batteries. Some SportsArt 9500 HR trainers, a kind of elliptical machine, can be modified to generate about 75 to 100 watts per hour, leading to a net zero energy usage.

Exercise clubs and gyms often have high energy needs. They have heating, air conditioning, security, laundry, water heating, and equipment energy costs. They compete with each other for members by the fees they charge their clients. As energy costs increase, club owners are looking for ways to lower costs and increase members and profits through alternative energy sources. Using the energy provided by their members is one way. Like other businesses they are also using solar energy.

Piezoelectric Sources

In 1880, brothers Jacques and Pierre Curie discovered the piezoelectric effect. This effect creates electricity when mechanical stress is applied to crystalline structures like tourmaline, quartz, topaz, ceramics, and cane sugars. The amount of electricity produced was proportional to the mechanical stress. First, ultrasonic transducers and quartz timing pieces were produced using this type of electrical energy generation. Currently, most airbags in cars use piezoelectric energy generation to inflate on impact.

There is a great deal of ongoing research on the use of piezoelectric energy. Medical research is investigating it as a source of energy for medical robots. Australian medical research scientists are developing miniature medical robots that use piezoelectrical energy. French researchers are developing materials that generate electrical power from the impact of rain. The Japanese are making mass transit station floors out of piezoelectric materials to generate power. The force of people, a mechanical stressor, on the crystals imbedded in the station floor creates voltage. In Britain, highways are being imbedded with power-generating piezoelectric crystals. It is estimated that 1 kilometer of road can generate 400 kilowatts of energy.

In terms of sustainability, piezoelectric energy ties energy production to mechanical use. The research is at such an early stage that it is unclear whether it will affect systems of nature in a way that impairs future generations. It does offer an alternative to the use of nonrenewable energy sources for electrical energy generation.

Energy Distribution

The concept of sustainable energy is more than its source. It also includes how that energy is distributed because that has an impact on systems of nature on which present and future life depends. Centralized sources of energy, such as nuclear fusion plants and nuclear reactors, have to be distributed to points of use. Most energy sources are converted at some point into electrical energy for distribution and use. Currently, electrical grids, substations, and lines carry the electricity to the point of use. Generally, the more centralized a source of energy, the more impact it has in its distribution. Electrical wires may create electrical magnetic waves that could disrupt life. Also, the current method of distributing electrical energy loses electricity because of resistance in the line and the transmission. Sustainability, and energy as now conceptualized, could be dramatically different if the wireless transmission of electrical energy becomes a reality. Smaller, localized sources of energy, such as rooftop windmills and solar panels, may be more sustainable because the power is created near the point of use.

Another aspect in the sustainability of energy other than its source and transmission is the environmental impact of the products used in its transmission and creation. Hydroenergy dependent on large dams may produce electrical power but may destroy whole watersheds and aquatic ecosystems. Solar panels with short lifespans and made with toxic materials may fill up landfills and leach into underground water supplies, eventually affecting the land. Nuclear fission energy sources create radioactive wastes that threaten entire ecosystems for tens of thousands of years. A comparison of energy sources requires full cost accounting of the product, its source, and how it is transmitted to the point of use.

References

Smil, Vaclav. 2008. *Energy in Nature and Society: General Energetics of Complex Systems.* Cambridge, MA: MIT Press.

Weiss, Charles, and William B. Bonvillian. 2009. *Structuring an Energy Technology Revolution.* Cambridge, MA: MIT Press.

FREE TRADE

Trade has brought development and wealth to many nations, particularly those nations that benefited from colonialism. Nations and people who became colonial commodities continue to endure the hardships of poverty and deprivation, even though some of these countries remain rich in natural resources, including undeveloped energy resources. Trade can create wealth, but it does not necessarily create wealthy communities. Wealth is not necessarily distributed broadly across a society by trade alone. Nor does trade necessarily support environmental respect or restoration. The challenge of the free trade movement in terms of sustainability is to show that its tremendous power can also achieve sustainable goals.

Labor advocates are often critical of free trade agreements because they do not protect the wage- and benefit-earning conditions of those workers in domestic markets against those wages and benefits of other workers off shore. Environmentalists are often suspicious of free trade agreements because they fail to provide environmental protection in off-shore areas that is as beneficial as that provided domestically.

These debates often create a dialogue of jobs versus environment, sometimes called job blackmail. When the choices offered are between unfair wages and damaging the environment, these are not good choices in terms of sustainability. Activists term them unacceptable choices or blackmail.

Can free trade be infused with values that create sustainable options for the local ecosystems and local economies? That is the challenge of sustainable free trade. ***See also* Volume 2, Chapter 2: Market-Based Strategies.**

GREEN CONSUMERISM

In developed nations, mass consumption of resources and energy to make products for consumer, household, and industrial uses has become a lifestyle and culture, as well as the basis for a consumer economy. The environmental footprint that this creates for developed nations vastly exceeds the footprint for less developed nations. The disproportionate use of resources and contribution to climate change of already wealthy nations far exceed the shares of those of developing nations. It is estimated that if everyone in the world were to attain contemporary U.S. standards of living, we would need several more planets as rich and hospitable as Earth to support life.

One response among developed, market-based nations is to create products and to advertise the use of environmentally friendly products and production methods. These emphasize lower impact on the world's ecosystems, and some also appeal to consciousness of global inequity. These products and services cost more, but mindful consumers seem willing to pay more for things that do not increase the damage to ecosystems and that assist development in poor countries and communities. The phenomenon of such consumption is also called "green consumerism."

Some have questioned whether green consumerism can reduce or eliminate overconsumption in developed countries. There are many questions about this kind of consuming, including whether the advertised goods are genuinely less wasteful or polluting when the entire cost of production and product lifecycle is considered. In addition, there is suspicion that if the problem is overconsumption as a cultural and moral value, encouraging more consumption is not going to lower impacts and environmental footprints in consumerist cultures. The argument is that, although it may have some small impact on individuals and businesses, it cannot fundamentally change an economy that pollutes and creates waste on such a large scale. This leads others to argue that, in developed countries the best solution to overconsumption is a change in fundamental values. For example, redefining happiness in less materialistic terms, and addressing intangibles bases for satisfaction such as security, opportunity for meaningful employment, and education. ***See also*** **Volume 2, Chapter 2: Market-Based Strategies.**

GREENWASHING

Sustainability is being touted in advertising of all sorts, from elite, non-profit sources like academic institutions, to cleaning products. Use and overuse of the term *sustainability* in advertising products, services, and programs have led to suspicion that the term means nothing, or is being used to mean anything. However, sustainability remains a popular key value across many stakeholder groups and individuals. It is a cause that seems unimpeachable to many people.

Businesses tend to note causes when they affect profits. Causes can and do affect business profits. There is rapid growth in using environmentally conscious proactive marketing of goods and services, which means to many that "causes" have gone mainstream. If two products are equal and one has a social or environmental cause, that is the product most people will purchase. Consumers have a much more positive opinion of the business if the business supports social causes. Business commitment to a cause is evidenced beyond consumers to employees, investors, and others. Business practices per se are considered part of purchasing influence. Consumers can also become activists if businesses are perceived as going too far in their environmental impacts and other community activities. With the modern ability to monitor environmental impacts, consumers can get environmental evidence quickly and accurately. Consumers and employees want businesses to engage in environmental causes. The pressure to label a practice or product as "sustainable" is very great for these factors.

In consumer products, the term *sustainable* is often used without any verifiable reference point. Similar terms such as *organic* often refer to verification by outside certifying organizations. A few organizations monitor and certify a range of standards important to sustainability such as LEED and ISO 14000 Tilt is a nonprofit organization in Oregon that sets organic standards. (see discussion in chapter 2 of this volume.). Other membership organizations require their members to account publicly for a range of behaviors related to sustainability in an annual report, such as the Ceres organizations. Other organizations aggregate a broad variety of rankings and publish their own sustainability indices.

The term *sustainability* has at least three different usages that are incommensurable, leading to charges of misuse or "green washing." The key difference in the use of this term is what the user means with respect to changing behaviors.

In one sense, the term *sustainability* means simply the ability to continue to do the same thing without changing behaviors, for example, questioning whether a company's profits are sustainable. In an environmental or ecological context, sustainability means the ability to achieve specific results without damaging the natural environment. Some companies have internalized this message as a second bottom line, profits and consequences to the natural environment. In the context of the original definition of sustainable development, it means redefining prosperity and happiness to accommodate the consequences to both the environment and community.

When some people use the term *sustainable* or *sustainability*, they are not including the concepts of environmental health, fairness, or justice, which makes the use of the term disingenuous to others who define sustainability as integrally related to one or both of these concepts.

Reference

Parr, Adrian. 2009. *Hijacking Sustainability*. New York: Columbia University Press.

Uncertainty Squared: The Challenge of Sustainability

Education at all levels focuses on teaching what we know and what we are reasonably sure about based on facts and inferences. At the higher and more abstract levels, education attempts to teach ways to think about what we are less sure of when facts are uncertain and inferences are attenuated. That is one level of uncertainty. When faced with uncertainty at this level, education responds with philosophies and values-based discussions that can be charted and certified by degree-granting programs. A different experience of uncertainty occurs when we do not know what we do not know. At that level of uncertainty, traditional education has had difficulty in conferring degrees and pedigrees. Many of the issues involved in the study of sustainability involve this second level of uncertainty. Beyond conflicting and competing values invoked by topics such as those noted in the Controversies section of each volume, the complexity of factual data that comes from each place studied make education a moving target, crossing many academic terrains, and occasionally requiring the honest admission of ignorance and humility.

The ultimate challenge for education in this situation is how to teach about complexity and ontological and epistemological ignorance. The ultimate challenge for a society struggling with issues of sustainability is how to make decisions under these circumstances.

Some meta-disciplines that long have struggled with these levels of uncertainty and the need for decision making. One such discipline is law. Law confronts the need for certainty in facts with a procedural acknowledgment that fact is socially constructed. In this discipline, fact-finding is done by a jury of social peers, recognizing that their conclusions are reflective of the community's own cultural perspectives, not necessarily a scientific enterprise. They are often instructed in how to resolve uncertainty in terms of a spectrum of probabilities ranging from a mere preponderance to a moral certainty beyond reasonable doubt.

When asked to make policy recommendations, scientists may experience the discomfort of approaching disciplinary limits in the form of uncertainty and the political consequences of uncertainty. Theories such as post-normal science attempt to place traditional sciences within a matrix of other knowledge such as community experience for the purpose of escaping the paralysis of judgment that uncertainty can impose within traditional limits.

These multidisciplinary, meta-disciplinary approaches to the problems of uncertainty are not yet comfortably housed within the architecture of contemporary educational structures, challenging educational institutions to come up with new structures.

INFORMATION AND KNOWLEDGE: ESSENTIAL FOR SUSTAINABILITY

The current state of knowledge about the environment is not adequate to know enough information for purposes of sustainability. As more information is learned, our knowledge of past and present environmental interactions and impacts increases. With this increase in knowledge comes increasing population growth and greater environmental impacts. A big question is whether knowledge growth can occur fast enough to mitigate the environmental impacts.

As the number and depth of sustainability stakeholders increase, the information available also can increase. One key concern is how "transparent" this environmental information is to all stakeholders. Is the information actually available and understandable to all stakeholders? This often depends on whether stakeholders have a right to the information. If they do not, they may not have access to it and it is not transparent. Adequate, accurate, and accessible environmental information is also necessary for meaningful participation.

Another constraint is the lack of important indicators of sustainable policy. Development of sustainability measures and indicators is one of the first steps in most public policy in sustainable development. Adequate, accurate, and accessible environmental information is a prerequisite for the development of sustainability indicators.

International Perspectives on Information for Sustainability

The main international approaches to developing information necessary for sustainability comes from Principle 10 of the Rio Declaration and Agenda 21, Section III:

> At the national level, each individual shall appropriate access to information concerning the environment that is held by public authorities, including information on hazardous materials and

> activities in their communities, and the opportunity to participate in the decision making process. States shall facilitate and encourage public awareness and participation by making information widely available.

Agenda 21 was adopted at the Earth Summit to implement the Rio Declaration. Chapter 40 of Agenda 21, section III focuses on information for decision making. It states:

> in sustainable development, everyone is a user and provider of information.

Many international standards on access to environmental information adopt U.S. standards found in the National Environmental Policy Act, the federal Administrative Procedure Act, and the Freedom of Information Act. All these laws have limitations in terms of real access to timely and accurate data. They are ultimately enforced in the courts, and access to the courts can be expensive and slow. Most of the environmental information developed by these agencies is for experts only and hard for the new stakeholders of sustainability, like the community, to understand. International standards also incorporate environmental information access standards of the Aarhus Convention in Europe. These international standards have little effective force in law.

Enormous advances in communication technology such as the Internet, however, now allow environmental information to be sent to a large number of previously excluded stakeholders. The technological ability to view any and all environments in real time may move nations to rapid development of sustainability indicators. Industries that seek out poor nations to pollute may not be able to hide as readily. Nations that exploit indigenous peoples for natural resource extraction will not be able to hide these acts of environmental degradation.

The Future Needs of Sustainability for Environmental Information

The development of sustainability indicators need environmental indicators on ecosystems and their relationship to global systems of nature on which present and future life depends. Ecosystem dimensions such as dynamic chemical and physical conditions, inventories of the plants and animals, and the actual impacts of all human use need be to accurate and accessible for the development of useful sustainable indicators. In-depth analysis of ecosystems is necessary. Ecosystem dimensions of food, fiber, water quantity and quality, carbon footprints and storage, soil conditions and shoreline changes, and biomass production and rates of use are necessary.

Environmental information should be available freely, quickly, and accurately. It needs to be monitored all the time, in real time. The concept of environmental information needs to be expanded to include the public health of all communities. Information about the state of a sustainable environment ultimately needs to be considered in all facets of life, in business, at home, and in other areas. Citizens need to have the capacity to develop and understand environmental information, and this is the role of education and civic engagement.

These changes will be expensive, and it is unsettled which stakeholder bears the cost. These changes may challenge current political and economic systems. Nonetheless, rapid advances in technology and environmental concern are pushing the development of accurate and accessible indicators of sustainability.

References

Felleman, John. 1997. *Deep Information: The Role of Information Policy in Environmental Sustainability.* Westport, CT: Greenwood Press.

Florin, Daniel J., 2006. *The New Environmental Regulation.* New York: Columbia University Press.

LABOR

The cost of paying people who work can be a major part of determining profitability. At times, those costs were treated in the same way as machinery—purchase and maintenance costs. This was the model of slavery and to a lesser extent share cropping. Both slavery and sharecropping or feudalism exists today, but most developed countries pay workers a regular wage for their labor.

Big differences exist between countries in terms of what wages are paid. Wages in developed countries are far higher than wages paid in less developed countries. Some businesses are able to move to areas where wages are less because they are service-oriented businesses, such as call centers or on-line services. Others can afford to move to cheaper labor markets because the cost of transportation of their products is less than the cost of labor. This practice is often called "outsourcing." Outsourcing jobs in a variety of industrial sectors makes economic sense only so long as the cost of transportation remains cheaper than the cost of labor. As the cost of oil and gas rose steeply, during 2007–2008, the cost of transporting goods became more expensive than the costs of paying domestic wages. If this dynamic were to remain constant over a period of time, it could have dramatic impacts on globalization in several manufacturing and producing sectors. Changing the relative costs of transportation and labor could fundamentally change globalization to localization. This is the fundamental nature of labor as a commodity in a global marketplace.

The costs and payment of a fair living wage to workers, regardless of where they are located in the world, are an important part of sustainability of economic enterprise. The need to earn a living to support oneself and families will drive individuals toward unsustainable behaviors in order to survive. In communities, we are joined together—businesses, workers, and our collective families—with our ecosystems in a relationship of survival. The choice of whether it is sustainable or merely short-term survivable turns instrumentally on our economic systems.

Labor and Transportation Costs in a Global Market

As businesses search for ways to increase their profits, they closely examine and compare the costs of labor and transportation. The choices that businesses make in these areas dramatically affect communities, the environment, and human health. These two costs can greatly affect the profit made on finished products. In addition, these two costs relate to each other in the global marketplace. Low labor costs in one country may tempt companies to outsource their manufacturing and service sector businesses to take advantage of these labor savings. Outsourcing of jobs can create hardships for families and communities, sometimes blighting once thriving neighborhoods and cities.

Savings in labor may be lost, however, if the cost of transporting goods to and from low-cost labor locations exceeds the cost savings on labor. As the price of petroleum fuels rise, the cost of transportation may well exceed the cost of labor, making local labor sources attractive, even if they are more expensive. For example, if the cost of labor in one country is much lower than in another country, the incentive to manufacture there is only as great as the relative costs of transporting goods. Any savings on labor costs will be spent on the costs of transportation. Decreasing fuel expenditures may help shrink the carbon footprint of a business. Localization helps to restore local communities.

What makes local sources of labor more expensive is often not simply the hourly wage, but also employer commitments to provide benefits such as health insurance, unemployment insurance, and retirement contributions. These commitments can be voluntarily negotiated by the employer and employees, or mandated by government. Outsourcing labor is a way for employers to avoid these commitments. Expensive transportation costs may exceed the costs of these social benefits, especially if these benefits are subsidized by government in the form of tax breaks to businesses that provide them.

Another way that labor costs can become global issues is through labor immigration. Some business sectors cannot outsource their high labor costs. For example, agriculture and construction labor must be done in a specific place, and laborers must come to those sites. In these sectors, labor savings are often achieved by hiring immigrant labor present in the country on work permits, or completely undocumented. These laborers may be willing to work for less, and without benefits because they do not know their rights, do not expect benefits, or will accept less because it is so much more than they can earn in their country of origin. In addition to foregoing wage and benefit protection, these labor forces are also less likely to insist on environmental and work safety measures. If they are present illegally, they may fear punishment and deportation more than the environmental and work safety hazards they encounter at work. Even apart from those sanctions, they may fear the loss of work more than these less tangible hazards. Therefore, this type of labor force also increases the likelihood of environmental and human health harms. In some areas, and for some crops, the yield is so substantial that it cannot be processed at harvest using only locally available legal laborers.

LOCALIZATION

The rise of globalization has also brought a rise in the movement of localization, aptly captured in the popular saying, "Think globally, act locally." The movement toward localization seeks to use local production and retailers to decrease the environmental impacts of goods and services. By decreasing the environmental footprint of goods and services, it is considered part of sustainable development. Another factor is that many small local businesses have a difficult time competing with national and multinational corporations. Large corporations can afford to undercut the price of the goods and services sold by local business people and drive them out of business unless the local business has another way to compete. Developing niche markets, increasing customer service, and increasing the image as a good neighbor are all ways small businesses try to compete. Promoting an image as environmentally sensitive is yet another, and localist movements do this. ***See also*** **Volume 2, Chapter 2: Globalization.**

Reference

Hess, David J., 2009. *Localist Movements in a Global Economy: Sustainability, Justice, and Urban Development in the US.* Cambridge, MA: MIT Press.

MARKET-BASED SOLUTIONS TO ENVIRONMENTAL DEGRADATION

Market-based strategies to achieve sustainability use conventional trading incentives to achieve goals related to sustainability. Buying and selling objects of trade are at the core of markets. Subsidies are payments that reward conduct deemed desirable. Taxes require payments from individuals or businesses. One use of taxes is as disincentives for conduct that government wants to slow or eliminate. The challenge that such solutions must face is how to incorporate environmental and social consequences that are not quantifiable in current marketplace terms.

Some people believe that markets can imitate nature successfully if they close the loops of productivity in ways that imitate natural closed loops. Using various tools of law and government, they argue that the power of the marketplace can reshape our economies to restore natural systems.

Market-based strategies often overlook the problems of poverty and poor communities because they do not participate effectively in marketplaces. The unintended impacts on poor people and communities can often undo the environmental good that such strategies intend. These effects are ignored when poor people and their communities are excluded from dialogue leading to public policy. This exclusion and these consequences lead to critiques by environmental justice advocates and other civil society groups of market-based approaches to solve problems of sustainability. ***See also* Volume 2, Chapter 2: Market-Based Strategies.**

References

Esty, Daniel C. and Andrew S. Winston. 2006. *Green to Gold: How Smart Companies Use Environmental Strategy to Innovate, Create Value, and Build Competitive Advantage.* New Haven, CT: Yale University Press.

Kraft, Michael E., and Sheldon Kamieniecki, eds. 2007. *Business and Environmental Policy: Corporate Interests in the American Political System.* New York: Columbia University Press.

Pollution and Emissions Trading

Companies are assigned the right to emit a certain amount of pollution by their permits. If these rights were made transferable to others, then buying and selling these rights to pollute could create a market in them. Such rights might be an incentive to companies to reduce pollution below permitted levels.

The net effect of such a market exchange would not necessarily reduce overall inventories of pollution unless coupled with a plan to reduce permitted limits. These are sometimes called "caps." In addition, the effect of transferability may be to concentrate more pollution into already burdened communities. The challenge that such solutions must face is how to incorporate environmental and social consequences that are not quantifiable in current marketplace terms.

The use of market-based solutions for sustainability relies heavily on principles of capitalism. These are discussed elsewhere in this section. Market-based solutions require something to sell to willing, able, and knowledgeable buyers. It could be that a sustainable lifestyle itself is a marketable commodity. Communities with sustainable schools, jobs, and homes may be cheaper, cleaner, and sustainably developed and create demand for such a lifestyle. Social concerns about the environmental impact of products such as cosmetics, clothes, building materials, and food already create a market demand for them if they are advertised as sustainable.

One example of how current market forces are applied to natural resource extraction is how Yvon Chouinard invests in logging. He was the founder of Patagonia, a company that prides itself in its environmental ethic and search for sustainability. Chouinard invested in a unique organization called Ecotrust Forests LLC. This fund is run by the conservation group called Ecotrust. Its approach is to buy 1,000- to 5,000-acre land parcels with 20-year-old forests. It then finds local loggers to harvest in a sustainable manner. This means taking out a few trees every year as they reach the appropriate size. It means harvesting them in ways that do not degrade the environment by respecting the canopy, the riparian areas, and the habitat of the native species. This system translates into harvesting no more than 25 to 35 percent of the trees each year and replanting enough trees to replace those that were harvested. Ecotrust hires wildlife biologists who examine the integrity of the ecosystem. Ecotrust retains ownership of the land. This is in contrast to traditional timber harvesting methods. Traditionally, large timber investment groups buy very large land parcels with 38- to 40-year-old trees. They prefer to clear cut them. This destroys all canopies, and can cause erosion and siltification. They then sell the land for whatever they can get for it. Ecotrust owned about 13,000 acres in 2008 and hopes to own about 250,000 acres by 2014. From 2006 to 2008, Ecotrust Forests LCC has made pretax returns of about 9 percent.

Another example of using market forces to achieve sustainability is winemaking in the Willamette Valley, Oregon. Salem, Oregon is the location of Low Input Viticulture and Enology (LIVE), which evaluates and certifies vineyards based on ecologically proven viticulture methods. About 33 percent of the approximately 250 wineries in the Willamette Valley are trying to incorporate sustainability. Some have used biodynamic farming approaches since the early 1990s. Others have built their physical infrastructure from recycled building materials. Still others try to use renewable energy sources such as solar and wind. The environment of the Willamette Valley does not have many diseases or insect problems so that the usual problem of pesticides did not occur. As approaches to sustainability developed in the general viticulture industry over time, they were easily applied to the Willamette Valley.

By treating a natural resource like renewable resource it is possible to use market forces to achieve sustainability. Sometimes it requires a radical change in approaches, as in logging. Other times it can mean working closely with systems of nature that are present when the industry begins, as in winemaking.

The challenge that such solutions must face is how to incorporate environmental and social consequences that are not quantifiable in current marketplace terms.

Reference

Fiorino, Daniel J., 2006. *The New Environmental Regulation.* New York: Columbia University Press.

POVERTY

Poverty is increasing at a greater rate than wealth. As a direct cause of the way natural resources are valued and used in our form of industrialized capitalism, the gap between rich and poor has widened as the systems of life have eroded. Poverty is related to human-driven degradation of our environment in several ways. Poverty intensifies the need for development in order to meet the basic needs of humankind. Where that developmental pressure is urgent, it may take place in ways that do damage to the surrounding ecological systems. In addition, poverty intensifies the vulnerability of human communities to working conditions that ultimately damage the webs of life. If workers are offered the choice between unhealthy work and deprivation for themselves or their families, they will choose unhealthy working conditions. This has been called "job blackmail."

Poverty poses many challenges to sustainability: can we be fair to future generations without being fair to our contemporaries with whom we share the webs of life on which all generations depend? Can we provide for a greater number of people within the limits of our ecosystems, especially if the basic human necessities are not within their means? Can we support a standard of living for a few that rests on a pattern of waste and pollution at the expense of life and health of all?

The Millennium Ecosystem Assessment (MA) established the relationship of poverty to environmental degradation in 2005. This finding supported the resolutions of the Millennium Development Goals (MDG) to eradicate poverty by 2015. The MA findings about the degree of ecosystem degradation concluded that the Development Goals would be harder to achieve in that timeframe.

Business as a sector of the economy has been slow to embrace the goals of eradication of poverty and other MDG goals. Businesses have come to appreciate the financial utility of eliminating waste and pollution from their production methods and products, partly led by examples of other businesses.

As a sector, business has been slower to recognize and embrace its role and responsibility for eradicating poverty. In his autobiography, Henry Ford recognized the business case for eradicating poverty. When he pioneered the industrialization of automobile manufacturing, he adopted the goal that the people who made his cars should be able to afford to buy them. In that way, he could ensure a ready market for the products he made. The income and employment that businesses provide ensure the social stability that supports all commercial activity. In much the same way that poverty degrades ecological systems of support, poverty also undermines the social conditions of peace and security on which our economy relies. Academics, nongovernmental organizations, governments, and others are now advocating a more conscious, intentional role for business in strategies of poverty eradication. Often, that role comes under the label of corporate social responsibility.

Corporate social responsibility to the environment is a controversial issue. Should it be voluntary or required? Generally, corporate social responsibility covers a wide variety of concerns. It usually includes respect for human rights and compliance with all applicable laws. It requires monitoring of environmental impacts to ensure this compliance with human rights and environmental laws. This requires risk assessments of some type for potential and actual human rights and environmental violation. Input from important stakeholders is encouraged, and the processes are transparent. Standardized corporate social responsibility response plans help to improve the corporate response. ***See also* Volume 2, Chapter 1: Introduction and Overview, the Need for Economic Development: Poverty.**

Reference

Chichilnishy, Graciela, and Geoffrey Heal. 2000. *Environmental Markets. Equity and Efficiency.* New York: Columbia University Press.

PRIVATE PROPERTY: CAN IT CONTINUE SUSTAINABLY?

Governments recognize and protect the idea of private, individual ownership of property to varying extents. To the extent that government or culture recognizes individual rights in land and other property, ownership gives individual owners the power to determine what choices to make regarding that property. In some places, this power is almost absolute, subordinating every other concern. The more absolute these rights of private property are thought to be, the more an owner's choices may be exercised without knowledge of, or regard for, the consequences of that choice for the ecology of the place or others within that community. To the extent that government ensures such absolute expectations of rights to property, sustainable choices for environment and for community may be compromised or wholly frustrated. When other interests contradict the interest of a private property owner, controversy is certain to occur.

Sustainability may contradict absolutist views of the rights of private property owners by asserting that interests in the environment and the community's interest in equity deserve equal consideration in decision making. This tension is evident in many controversies regarding waste, pollution, and inequities resulting from human uses of resources like land and water.

Owners of private property often do not wish to be held accountable for the consequences of their choices about waste or pollution that they generate, and that can be (or historically have been) passed along to others, including neighbors, the environment, or distant places and strangers. Economists call the passing of these costs along to someone else "externalities."

Advocates of sustainability insist that those whose actions create waste and pollution should be responsible for the externalities they cause. They advocate mechanisms forcing the generator of waste or pollution to pay the full and true costs of these acts to others, and to communities. Mechanisms forcing producers to pay these costs include taxes or fees. To the extent that a producer pays for these "downstream" consequences, they are said to be internalized. In addition, to the extent that producers wish to avoid these costs, they will have an incentive to take measures resulting in more sustainable behaviors, less waste, and less pollution. ***See also* Volume 2, Chapter 3: Barriers to Governmental Roles in Sustainability.**

References

Freyfogle, Eric T. 2003. *The Land We Share: Private Property and the Common Good.* Washington, DC: Island Press/Shearwater Books.

Fuchs, D. A. 2003. *An Institutional Basis for Environmental Stewardship: The Structure and Quality of Property Rights.* New York: Springer.

Geisler, Charles, and Gail Daneker, eds. 1997. *Property and Values: Alternatives to Public and Private Ownership.* Washington, DC: Island Press.

REAL ESTATE

Real estate is a legal conceptualization of land that makes it into a commodity that can be bought and sold. It applies equally to all types of land without regard to its environmental or ecological significance. It does not impart any duties to future generations, to any member of the public other than buyer and seller, and to the environment. Real state basically straps a market-based concept onto a limited and necessary public resource—land. Real estate is intrinsic to private property models, but it relies on large direct and indirect public subsidies to operate. For example, the rate of interest on many mortgages is a large tax deduction. Most mortgages require ownership of private property. Real property is in contrast to personal property. The latter can be moved, but real property cannot be moved.

Real estate interests in private property can be divided into various legal interests such as easements. Some environmental conservation efforts have focused on preserving important aspects of the environment with the use of conservation easements. Easements are few in number and difficult to enforce. ***See also* Volume 2, Chapter 3: Barriers to Governmental Roles in Sustainability.**

References

Angotti, Tom. 2008. *New York for Sale: Community Planning Confronts Global Real Estate.* Cambridge, MA: MIT Press.

Fairfax, Sally K. et al. 2005. *Buying Nature: The Limits of Land Acquisition as a Conservation Strategy, 1780–2004.* Cambridge, MA: MIT Press.

RESOURCE SUBSIDIES

Governments have long used subsidies to encourage certain economically desirable practices by paying companies to engage in those practices. Subsidies have been offered to businesses in various forms including tax breaks and direct payments. In agricultural policy, subsidies have become a long-term tool to manage everything from land and water conservation to crop rotation. Especially in agriculture, subsidies have become an expectation bordering on a right. Farmers in some countries rely on subsidies to make a living from their land, rather than on the income from production. Agricultural subsidies are a major feature affecting the price of commodities in international trade. In some countries, the success of these agricultural subsidies has transformed farmers from a sector focused on the food security of their nation to international commodity brokers.

> When subsidies are used to encourage sustainable behaviors and practices, they can be an important economic tool in the short term. Subsidies are an important way to manage costs and disincentives associated with a transitional period. While a business is making a change to another way of doing business, or another technology, losses are often probable. To make those losses manageable until the change is completed, subsidies may be a way of encouraging necessary change. However, when subsidies become a long-term accommodation or sinecure, true and full costs are no longer being internalized by the business, and are instead imposed externally on other constituencies. In the long term, this enables unsustainable, harmful behaviors.

Constituencies now accustomed to receiving subsidies will resist their loss, sometimes violently protesting their removal. Efforts by the WTO to eliminate agricultural subsidies in international trade, so that agricultural commodities would be traded internationally without distortion from these sources, have floundered repeatedly. These discussions have taken place most recently in the Doha Round of WTO talks. The WTO is an association of member nations who have agreed to eliminate trade barriers between themselves. It is the largest free trade agreement in the world. It was known as the General Agreement on Tariffs and Trade until 1995. The process of negotiations about different types of trade barriers is a continuing process. Negotiations are known as "Rounds" and each Round is called by the name of the city in which it took place.

At the Doha Round, developed countries represented by Europe and the United States of North America and developing nations represented by Brazil and India failed to reach an agreement for eliminating agricultural subsidies in commodities like cotton, corn, and rice. During negotiations, the prices of these commodities soared under the influence of international oil speculation. The negotiations failed to deal not only with the differences between developed countries and developing countries interests, but also with the effects of removing subsidies on farmers and agribusiness employees.

Some economic sectors are not sustainable as currently configured. The future of fossil fuels in a sustainable economy means that workers in those areas such as oil workers, coal miners, and others are facing an uncertain future. Similarly, agricultural workers in crops heavily subsidized by government, forest workers, and their communities are often subsidized by the government. Without subsidies, these workers and their communities are threatened with economic extinction. This threat often pits their economic interests against the environment that supports them. Without a plan for a fair transition to clean production or sustainable practices, businesses and their employees will suffer. The planning for such a transition can be driven by moral and ethical considerations, or by governmental intervention.

Energy Subsidies for Unsustainable Incumbent Energy Sources

Most of the incumbent energy sources of oil and gas contribute more than 90 percent to the world's greenhouse gases that are causing global warming and pose threats to systems of nature on which future life depends. The concern is that by subsidizing these activities, the rate of conversion to more sustainable energy sources is slower, and the ability to slow down global warming is slower. Others have concerns that the free market would not allow government subsidies. If these industries are given special treatment and subsidies, they can act more wastefully and pollute more instead of more efficiently and internalizing the costs of their production. Many of these industries are heavily subsidized so that governments interested in economic development and growth can attract cheap energy sources. Cheap energy attracts economic development because it allows other industries to increase profits. Renewable energy sources also receive subsidies, but at far fewer amounts. They also tend to have far fewer environmental impacts than nonrenewable energy sources.

Subsidies: Acceptable Government Intervention in a "Free" Market?

Subsidies represent a contradiction in traditional theories of capitalism of a free market without government intervention. Subsidies are a government intervention in that they are a transfer of economic resources to market participants that affect either prices or production costs. These transfers of costs can be direct and many are indirect. Shifting the risks of a private market from the industry to the government is one indirect market transfer. For example, the government can underwrite loans the industry makes for research and exploration. The government can also allow tax deductions for these activities. The government can also allow tax deductions for environmental fines and penalties, as it did for the penalties assessed for the Exxon Valdez spill. Government can also develop special programs that are for all businesses but happen to help a nonrenewable energy sector substantially more than others. Other indirect subsidies are government contracts that direct consumption of the nonrenewable energy source. These artificially inflate demand, such as using only gas and oil for the Department of Defense in times of war or otherwise. Subsidies can occur at all levels of government, from federal, state, and local levels. This allows energy industries to have governments compete with each other for economic development by offering different subsidy packages. This can also include tax relief that other industries and consumers pay. There are a range of costs engendered by the gas and oil industries that are not categorized as subsidies but cost the public in intangible ways, for example, drilling in wilderness and otherwise protected areas. The Arctic Wilderness Preserve is one example. Although it is not a direct subsidy, it is an activity that would not be otherwise allowed and that gives a private industry protection in an otherwise "free" market.

The range and nontransparency of these subsidies make their financial amounts difficult to determine with a high degree of certainty. It is estimated that the U.S. government subsidizes fossil fuels between $15 and $35 billion a year, even though many of these industries have made record high profits and the U.S. government is operating at record high deficits. These subsidies to oil and gas industries have not translated into economic growth for the nation or states.

Subsidies for Nuclear Energy

Behind gas and oil subsidies, nuclear energy receives the most subsidies. There are approximately 114 to 120 nuclear power plants in various operational phases in the United States. This energy resource would not be commercially viable without subsidies. In recent years the industry has quickly asserted itself as a subsidy recipient. If the issue of hazardous wastes is set aside, nuclear energy is cleaner than gas and oil energy sources. After a hiatus of two decades, the George W. Bush administration streamlined the permitting process and gave additional subsidies to these businesses. These permits allow them to override the concerns of state and local governments, which are considered an indirect subsidy by some. These permits last for 20 years and can be renewed easily for another 20 years.

Nuclear power does pose environmental risks because the problem of the hazardous wastes it produces has not been solved. These wastes are part of uranium enrichment and the application of plutonium as a fuel. After these elements are used, their wastes are lethal to all life for tens of thousands of years. The waste needs to go somewhere, which means wherever it goes, no life or natural systems on which future generations will rely can occur. As these wastes accumulate, absent any new developments that decreases their lethal hazardous characteristics, more natural systems will be affected. Given the length of time they are dangerous, there is a concern that a natural disaster such as an earthquake, flooding, fire, or large storm could destroy any waste containment site and spread the lethal hazardous nuclear waste into a broader spectrum of the environment.

References

Pope, Carl, and Robert Wages. 2000. "Green Growth: Agenda for a Just Transition to a Sustainable Economy." In *The Next Agenda: Blueprint For A New Progressive Movement,* eds. Robert L. Borosage, and Roger Hickey, 249–75. Boulder, CO: Westview Press.

Macfarlane, Allison M., and Rodney C. Ewing. 2006. *Uncertainty Underground: Yucca Mountain and the Nation's High-Level Nuclear Waste.* Cambridge, MA: MIT Press.

Meyers, Norman and Jennifer Kent. 2001. *Perverse Subsidies: How Tax Dollars Harm the Environment and Economy*. Washington, DC: Island Press.

Weiss, Charles, and William Bonvillian. 2009. *Structuring an Energy Technology Revolution.* New York: Columbia University Press.

RISK ASSESSMENT

Risk assessment is a form of analysis of the probability and magnitude of harm from various events and activities. It is widely used to make decisions. Insurance relies on this computation of risk of harm in making decisions about whether to insure, and if so, how much to charge for insurance. Risk assessment is also used in decisions about development projects, and it is used by governments in budgets and planning activities. Related to the science of risk assessment, risk management determines how to plan for and communicate about risks. Risk perception is a science devoted to examining the qualitative aspects of risk, not simply its quantitative aspects.

Risk assessment, as it used in business and industry, often refers to the risk to profit-making and to capital assets. Downside risk refers to the exposure a business has to profit loss. Upside risk refers to the risk a business has to increase profits. Business plans examining market risks frequently make reference to the regulatory environment as generally decreasing profit potential. The regulatory environment can include finance, insurance, investment, and environmental regulations. After the Great Depression of the 1930s, the U.S. government created regulations to separate finance, banking, and investment functions so that the boom and bust cycles of the market would not threaten the security of the nation. Beginning with President Reagan and ending with President George W. Bush, the government deregulated these industries, and the robust recession of 2008 began. Environmental regulations began in the 1970s in the United States and were fought by industry in courts, legislatures, and elections. With the deregulation of economic regulations came the deregulation of the environment, usually accompanied by lax enforcement. In 2008, for the first time the emissions of sulfur dioxide were increased, and the progress made in acid rain and reversing the death spiral of the Adirondack Lakes ceased. These and other environmental degradations do not bode well for the environment unless and until sustainable development is emphasized in government.

There are many controversial nuances to the assumption of environmental risks within the narrow, profit-focused, and short-term window of business risk. It may not be possible for any one business or industry to shoulder all the environmental risks it caused. The threat that businesses may be forced to do so may stop technological innovation that could be more sustainable. Others object to the privatization of profits for a few large businesses and corporations while their liabilities and losses are socialized by the government among low- and middle-income taxpayers. It is becoming increasingly clear that businesses and industries that do pollute are acting wastefully by overconsuming resources and not incorporating the true costs of their operations into their bottom line. It is also becoming increasingly clear they will continue to do so until natural resources for future generations are irreparably destroyed unless there is a strong push by government for sustainable development. Business and other organizations focusing on economic development will need to examine how environmental risks are measured and perceived.

Risk assessment is an alternative to cost-benefit analysis as a way to make decisions. It is a form of analysis of the probability and magnitude of harm from various events and activities. It is widely used to make decisions. Insurance relies on this computation of risk of harm in making decisions about whether to insure, and if so, how much to charge for insurance. Risk assessment is also used in decisions about development projects, and it is used by governments in budgets and planning activities.

Related to the science of risk assessment, risk management determines how to plan for and communicate about risks. Risk perception is a science devoted to examining the qualitative aspects of risk, not simply its quantitative aspects.

Risk assessment chooses what factors and events to consider. In the process of choosing, the assessment may be deliberately constructed to exclude certain factors. Often these choices reflect the desire to simplify the task, but the cost of oversimplification is the heightened risk of coming to erroneous decisions about management and communication.

Risk assessment is a complex and controversial method of decision making. Basic risk assessment follows a single chemical in a single pathway through air, water, or land. The choice of which chemical to monitor often depends on whether it is considered harmful to humans, and this is called a human health risk assessment. The choice of which chemicals to monitor is controversial for several reasons. One is that humans react to chemicals differently; the amount or dose of the chemical may determine if it is "harmful or "adverse." The dose response rate among the public in the United States to aspirin could be up to 1,000 times different. Another reason it is controversial is disagreement about the choice of which chemicals need monitoring. Pesticide manufacturers may claim it is safe, while consumer groups may claim it is not safe. The application of the precautionary principle used in sustainability to a single pathway human health risk assessment would ensure that the chemical does not harm natural systems on which future life depends. That would include present-day humans.

Aggregate risk assessment follows many chemicals through a single pathway through air, water, or land, and then adds the risk together. This is also known as the additive principle of risk assessment. It has the same dose response and chemical selection issues as human health risk assessments. Many chemicals interact with each other. Aggregate risk assessment can also mean multiple pathways; that is, it follows one or more chemicals through both air and water and sometimes land. Last, aggregate risk assessment can follow multiple chemicals through multiple pathways, but it always uses the additive principle of risk. Because of the potentially greater threat posed to natural systems on which future life depends, the application of the precautionary principle of sustainability is arguably greater.

Aggregate risk assessment is not cumulative risk assessment. Cumulative risk assessment considers all past, present, and future chemicals; their multiple pathways; and the endpoint of the chemicals at a given point in time. Many argue that only cumulative risk assessment is viable for sustainability. Others argue that it is an impossible task given current states of environmental evaluation and knowledge. Cumulative emissions measures all the chemicals that come from a given place or industrial process. Cumulative impacts measure the effects of all those emissions over time on a given endpoint, such as a human or a food product.

In 1998 the U.S. EPA finished its Cumulative Exposure Project based on the additive principle. In this project it modeled the concentrations of the hazardous air pollutants in 1990 all across the United States. It then combined these measurements with unit risk estimates to estimate the potential increase in cancer risk from multiple hazardous air pollutants. The cancer risks from each hazardous air pollutant were added across pollutants in each census tract to estimate a total cancer risk in each census tract. When uncertainties like synergistic interactions, underestimation of ambient concentrations of hazardous air pollutants, different human vulnerabilities, and changes to potency estimates were considered, cancer risk may have been underestimated by 15 percent.

Ecological risk assessment applies principles of assessing risk using the ecology as a chemical endpoint, as opposed to human health risk assessment that uses humans as the chemical endpoint. Ecological risk

Alberta Desert Sand Cumulative Risk Assessment Project in the Oilfields

In Canada, the concept of private property is more limited than in the United States, and provinces own the mineral, gas, and oil rights to their lands. In the north of the province of Alberta, there is oil located in a vast area of so-called tar sand about the size of Florida. This is also an area with biological diversity and indigenous peoples dependent on the environment for sustenance. The tar sands are located in one of the largest boreal forests in the world, valuable among other reasons for its role in carbon sequestration and as bird habitat.

To get to the oil without damaging the environment, some type of risk assessment was necessary. Canada decided to allow oil extraction from the tar sands if the oil companies managed all the risks including those that accumulated.

The province of Alberta, however, has allowed extraction on more than 65,000 kilometers of land without environmental assessment, which was not part of the earlier agreements. Many also consider the tar sands of Alberta to be a large source of greenhouse gases in Canada, emitting an estimated 40 tons of greenhouse gases annually, projected to increase to 141 tons by 2020. The type of oil production in these tar fields used large amounts of water, and many parts of this area are considered a desert. Furthermore, many believe that the tailings ponds, where waste from oil extraction is processed, are too close to the Athabasca River. They cover 130 square kilometers and may be decreasing biodiversity and causing cancer clusters among indigenous people.

Oil production in the desert sands of the boreal forests of northern Alberta presented a unique opportunity to study the management of cumulative impacts on an ecosystem. It demonstrated that the many decisions of a fragmented government approach in a context of an increasing climate change underestimate environmental and human impacts. This is an important lesson for other countries that have less governmental regulation, less accountability, and valuable natural resources.

assessment is currently mandated in the cleanup of U.S. Superfund sites on the National Priority List. (www.epa.gov/superfund/sites/npl/npl.htm Ecological risk assessment follows one of more chemicals through the entire ecology of an area, the land, air, water, flora, fauna, and all life forms. It often requires an endpoint in time because healthy ecosystems are dynamic. It also has controversies of dose response and choice of which chemicals to monitor. Also, the current state of knowledge about ecosystems is incomplete.

All the preceding risk assessment approaches do not usually include actual chemical actions. Most assume that the risk is simply the risk of one chemical added to the risk of another chemical. Most chemicals, however, have some effect on other chemicals, especially in cumulative and ecological assessment risk assessments. Chemicals can synergize, or increase the risk, of another chemical. Chemicals can also antagonize, or reduce the risk of another chemical. Chemicals can interact with their endpoint, whether it is human or ecosystem, to increase or decrease risks. Some postulate this chemical interaction with endpoints may be one reason there is such a wide variance with the dose response rates. Chemicals can also be simply inert with one another, or may serve as catalyst. A catalyst chemical may cause a reaction between chemicals but not be part of the final chemical product.

Risk assessment is a primary tool used to make current environmental decisions, but it is still unrefined, expensive, and controversial.

References

Heal, Geoffrey. 2008. *When Principles Pay: Corporate Social Responsibility and the Bottom Line.* New York: Columbia University Business School.

Robson, Mark G. and William E. Toscano, William E. 2007. *Risk Assessment for Environmental Health.* Hoboken, NJ: Jossey-Bass.

Townsend, Amy K. 2006. *Green Business: A Five Part Model for Creating an Environmentally Responsible Company.* Atglen, PA: Schiffer.

Whiteside, Kerry H. 2006. *Precautionary Politics: Principle and Practice in Confronting Environmental Risk.* Cambridge, MA: MIT Press.

Public Perception of Risk

To manage risk effectively, risk managers may rely on their power to enforce their decisions involuntarily or voluntary. Risk perception reveals how people feel about different types of risk. Sometimes these perceptions can seem surprising or even counterintuitive. Perceptions are real and help to determine the nature and quality of behavior when faced with certain types of risk. When certain perceptions are excluded from management decisions, there is heightened risk of noncompliance or resistance to management decisions. Decisions to discount or exclude certain perceptions are made based on lack of scientific expertise. This may result in popularly discredited management decisions. Sometimes decisions to discount or exclude certain risk perceptions are often made based on gender, race or economic class with the same results. Voluntary compliance based on commonly held perceptions of risk eliminates much of the need to use involuntary enforcement mechanisms.

Race and Gender Differences in Environmental Risk Perception: A Big Divide

How do perceptions of risk affect a stakeholder's involvement in environmental decisions? This question is controversial, but it will have to be fully addressed as social policies of sustainability become implemented. Even if the perception of environmental risk is not accurate, it can cause physiological impacts on the public. The perception of environmental risk or hazard to oneself or one's loved one creates stress that can lead to heart attacks, hypertension, and strokes.

A stakeholder is defined as a person who has a stake, or an interest, in the outcome of a decision. How a decision is framed can determine who the stakeholders are. In recent history, environmental decisions were framed as those between environmentalists and industry. Currently environmental decisions are framed to include a public health component, as the results of incomplete environmental decisions have begun to accumulate into more than just degrading ecosystems but to affecting public health. This enlarges the stakeholder pool to include communities, governments, environmentalists, and industry. Under a social policy of sustainability, the stakeholder groups may be even more expansive because of the inclusion of ecosystems and longer term environmental considerations.

The perception of risk may not be the actual risk posed; however, even scientific versions of risk posed may be inaccurate. Even expert risk assessors have been shown to have bias in their conclusions based on how they view the world. Fear of catastrophic risk, like a nuclear accident such as Three Mile Island, may pose little scientific risk but may have a large perception of risk. Dread risks, like cancer, may pose little scientific risk per million people, but the perception of dreaded diseases like cancer can increase the perception of risk. The perception of risk can often depend on whether a person is assigning risk for others or assigning risk for herself or loved ones. When assigning risks to environment or to ecosystems the perception of risk may be lower without humans than with humans. Under a social policy of sustainability, the propensity to lower risks in nonhuman environments is an issue.

With each new set of stakeholders, and sometimes within traditional stakeholder groups, the perception of risk from active and passive environmental decisions changes. An active environmental decision usually has a proponent with an idea for some type of project or activity that has an environmental impact. A passive environmental decision is one that allows the current way of environmental decision making to continue. Because many past and present environmental decision-making processes are incomplete and have significant environmental consequences, many passive environmental decision-making processes will be revisited under more comprehensive policies of sustainability. The perception of risk in both passive and active environmental decisions will be a significant factor in sustainability policies.

Race, gender, income, and educational level all change the perception of environmental risk one perceives and is willing to assign to others and to the environment. These perceptions are the result of many factors, such as a history of environmentally disproportionate risks. These perceptions of risk can also be the result of being a member of a privileged social group, whether or not that group acknowledges the privilege. As social policies of sustainability revisit passive environmental decisions and become more inclusive of formerly excluded stakeholder groups, some sustainability decisions can become controversial.

In some research examining 30 categories of risk perception, nonwhites perceived environmental risks as approximately 30 percent higher than whites. Females perceived environmental risks as about 30 percent higher than males. Some have argued this leaves a white male aura of invincibility. White males are a privileged group in U.S. society, and this may be an area of unacknowledged privilege. White males are a dominant group in many areas of traditional environmental decision making, such as the U.S. EPA and as chief executive officers of most industries. Some have argued that it is the lack of scientific training that accounts for the disparity between male and female perception of environmental risks. Research shows, however, that the perception of environmental risk between males and females is still about 30 percent different in samples of males and females with terminal degrees in scientific fields.

If white males made environmental decisions based on low perceptions of environmental risk, then many of these decisions may have to be reopened as policies of sustainability begin implementation. This may include many active and passive environmental decisions of the past. An open question is whether the application of the precautionary principle of sustainability, that a decision is postponed if it could harm natural systems on which future life depends, will reopen past environmental decisions.

References

Beer, Tom, and Alik Ismail-Zadeh. 2002. *Risk Science and Sustainability: Science for Reduction of Risk and Sustainable Development of Society.* New York: Springer.

Flynn, James, Paul C. Slovic, and K. Gender Mertz. "Race, and Perception of Environmental Health Risks." *Journal of Risk Analysis* 14 (1994):1101–8.

TECHNOLOGY TRANSFER, INTELLECTUAL PROPERTY RIGHTS, AND THE DEVELOPING WORLD

The developing world is host to the majority of what we do not know in terms of plants, species, and the material world. As we discover more, explore more, and come to appreciate the extraordinary richness of these often-impoverished places, we understand the vast repository of what we do not yet know. Sometimes we call this vast unknown epistemological ignorance, and we ask science or religion to fill in the blank spaces with answers that will serve us now. Others have found answers and hope in the discoveries they have made by exploration of new places and cultures.

Intellectual property law in developed countries was first created to provide artists and creative persons with incentives to keep on creating useful things by assuring creators a stream of income from their creations. Over time, this assurance created valuable property rights best exploited by publishers and other enterprises that distributed these works through media such as radio, television, computers, and other networks.

When explorers and discoverers find new valuable species in developing nations and among original peoples, they are often working for corporations that can finance such ventures, and in return these venturors expect the intellectual property rights to these finds. Indigenous peoples, however, often have different ideas of entitlement and no idea of the financial consequences of exploitation of their recent visitors.

References

Dillard, Jesse, Veronica Dujon, and Mary C. King. 2008. *Understanding the Social Dimension of Sustainability.* Hampshire, UK: Routledge.

Paarlberg, Robert. 2009. *Starved for Science: How Biotechnology Is Being Kept out of Africa.* Cambridge, MA: Harvard University Press.

TRAGEDY OF THE COMMONS

The tragedy of the commons is a story about overgrazing of commonly owned land. This famous narrative says that when people are allowed to graze their animals on land that belonged to everyone, they will inevitably graze too many animals on the land, causing it to become overgrazed. This story is often used to describe the way that complex ecosystems can collapse when left unplanned or managed for the common good. In a contemporary essay that revives the moral and economic debates about population, carrying capacity, and private ownership of land, Garret Hardin retells the fable.

> The tragedy of the commons develops in this way. Picture a pasture open to all. It is to be expected that each herdsman will try to keep as many cattle as possible on the commons. Such an arrangement may work reasonably satisfactorily for centuries because tribal wars, poaching, and disease keep the numbers of both man and beast well below the carrying capacity of the land. Finally, however, comes the day of reckoning, that is, the day when the long-desired goal of social stability becomes a reality. At this point, the inherent logic of the commons remorselessly generates tragedy.
>
> As a rational being, each herdsman seeks to maximize his gain. Explicitly or implicitly, more or less consciously, he asks, "What is the utility *to me* of adding one more animal to my herd?" This utility has one negative and one positive component.
>
> 1. The positive component is a function of the increment of one animal. Since the herdsman receives all the proceeds from the sale of the additional animal, the positive utility is nearly +1.
> 2. The negative component is a function of the additional overgrazing created by one more animal. Since, however, the effects of overgrazing are shared by all the herdsmen, the negative utility for any particular decision-making herdsman is only a fraction of 1.
>
> Adding together the component partial utilities, the rational herdsman concludes that the only sensible course for him to pursue is to add another animal to his herd. And another; and another. . . . But this is the conclusion reached by each and every rational herdsman sharing a commons. Therein is the tragedy. Each man is locked into a system

that compels him to increase his herd without limit—in a world that is limited. Ruin is the destination toward which all men rush, each pursuing his own best interest in a society that believes in the freedom of the commons. Freedom in a commons brings ruin to all.

From Hardin, G. (1968). Tragedy of the Commons. *Science,* 162, 1243–1248. The American Association for the Advancement of Science. Reprinted with permission from AAAS.

This narrative shapes many contemporary arguments about public lands and other resources that are not privately owned or valued. ***See also*** **Volume 1, Chapter 4: Tragedy of the Commons.**

References

Hardin, G. "Tragedy of the Commons." *Science* 162 (1968):1243–48. Online at www.sciencemag.org/cgi/content/full/162/3859/1243.

Manning, Robert E. 2007. *Parks and Carrying Capacity: Commons without a Tragedy.* Washington, DC: Island Press.

Prisoner's Dilemma

The Prisoner's Dilemma is another story told to illustrate patterns of human behavior when we are faced with choices that appear to maximize one person's advantage over another. In this story, four prisoners are given the choice to cooperate with officials holding them in exchange for treatment that is more favorable. If none of them cooperates, they will all go free. The best outcome for all of them is to remain silent, but self-interest may deceive them into a choice that is not in their best long-term interest.

One key to solving the prisoner's dilemma, as in the Tragedy of the Commons, is to overcome the impulse to short-term self-interest that results in fewer benefits overall. The dilemma and its tragic consequences for humans and nature can be resolved if the individuals involved can come to see both the folly of their immediate choices and the desirability of changing them in favor of behavior that benefits all of them, whether they are shepherds or prisoners.

Public Lands

Some publicly owned lands are open to managed uses for private benefits. Ranchers may lease public lands for grazing sheep and cattle. Oil and gas corporations lease the right to extract oil from public lands. Mining companies own the rights to mine some public lands for their mineral content. National forests can be logged for private profit. But other types of publicly owned lands are held immune from private profit to a greater extent. National parks are open for tourism, and concessionaires offer tourist-related services like hotels and food services, but they are not open for extractive purposes. These differences arise from very different public policies about conservation and environmentalism, as well as from different geographies and ecosystems.

There is growing tension between private uses that can degrade the environmental systems of an area, and the idea of areas that should not be used or developed at all. Particularly in forests and ocean areas, there is growing concern that private uses of resources in these areas have not been respectful of the interconnected ecosystems and habitat. Private users of public lands now expect that they may exploit the resources of publicly owned lands as a matter of right because they have been allowed to do so in the past. Managers of these lands are under increasing pressure to exclude private uses, even limiting tourism and related services to preserve the integrity of ecosystems for the benefit of nonhuman species. ***See also* Volume 1, Chapter 4: The Tragedy of the Commons.**

DEMOCRACY AND SUSTAINABILITY

Basic democratic theory requires the participation and consent of the many that are governed in public law and policy. Exclusion and lack of consent make government action illegitimate. For those who see the need for change toward sustainable behavior as urgent and widespread, the need to consult widely and to develop political constituencies and leadership for sustainable change seems too complex, indirect, and compromised to promise hope of effective change. Some have argued that timely change on the scale that we would need to avoid collapsing our webs of life can be achieved only by a different arrangement of power, such as a plutocracy or oligarchy of properly motivated people or organizations.

One possible response to this argument is that the time taken to ensure genuine participation on a local level also ensures compliance and eliminates the need for costly and slow enforcement mechanisms. This is one reason that Agenda 21 recommends that government action toward sustainability be formulated at the lowest effective level of government. Another response lies in the observation that effective action need not mobilize all of the population. Effective changes can be generated by smaller groups that exercise symbolic influence if not actual power. The power of the tipping point can be exercised voluntarily even in the absence of governmentally sanction power to coerce.

Reference

Hester, Randolph T. 2006. *Design for Ecological Democracy.* Cambridge, MA: MIT Press.

IMMIGRATION

Movement of people from one area to another is so ancient that it defines our human genome. These patterns of migration may be caused by environmental or political causes. Issues caused by these migrations may arise from displacement of people from land and livelihoods. In a contemporary sense, these migrations raise political issues when they cross-political jurisdictions and boundaries, which may have more to do with history than current ecological or social realities in the region.

In developed countries, migrations bring younger workers. If the countries have adopted a public retirement system and the expectation of a protected retirement, these younger workers will need to be able to earn high incomes to support the social system of retirement. Developed countries are not repopulating themselves at a rate or an educational level that can support these expectations. In developing countries, the challenge of creating an educated generation will be to keep them interested in remaining in their country of origin.

The challenge of immigration into developed countries is one of recognition, identity, and justice. Developed countries do not yet acknowledge their dependence on the contributions of young workers from other countries, especially countries composed of people of color who may have better educational credentials than young workers within these countries. The race, color, and religion of young immigrants may disturb the cultural identity of these countries. Moreover, laws ensuring wages, benefits, and security of tenure to these workers may undermine both social and economic assumptions.

Reference

Munier, Nolberto. 2007. *Handbook on Urban Sustainability*. New York: Springer.

GLOBALIZATION OF MARKETS

Can markets develop sustainably? Are the concepts inherent in markets, such as supply and demand, flexible enough to sustain us? Wealth creation does not necessarily instill values in favor of nature or communities. What does instill these values, and can it work on a global basis in the absence of a unified governmental force? One major way is through organizations such as the United Nations, the World Bank, the Union of Concerned Scientists, and the International Wildlife Federation requiring and developing concepts of sustainability. In the business world of developing free markets, these values emerge as soft international law in the form of trade limitations. Soft international law becomes internalized as international custom and practice. Sustainability as a value in practice becomes codified on a national basis. It then becomes hard international law underwritten by multilateral treaties such as the Stockholm Accords or the Montreal Protocols.

Globalization means the ever-increasing network of connections between individuals and organizations made possible by technology and transportation. These networks include information technologies, as well as other communications that facilitate the transfer of capital, goods, and redistribution of labor around the globe. Globalization can mean ever-widening markets, as governments agree to remove barriers to trade and immigration between them. Globalization can also mean the deep presence of markets through virtual technologies. In earlier centuries, markets were limited in time and locations. Today, market transactions can occur virtually anytime by more people, and anywhere. Markets have been good vehicles for increasing wealth, but growth in marketplaces has also accompanied alarming trends in the rate of climate change and other environmental deterioration, as well as growth in poverty. The rapid increase in natural resource use, the reliance on nonrenewable energy sources, the failure to include environmental impacts in the costs of production, the failure to include health impacts in the costs of productions, the lack of inclusion of cumulative environmental and health effects are all parts of market approaches that are hard to incorporate under most current models of sustainability.

Many information technologies have the potential to lessen our environmental impact. The rapid growth of information technologies could assist sustainability efforts in areas such as real-time monitoring and forecasting environmental conditions. The rise of social concern and public policy around sustainability itself drives new markets for sustainable products and processes.

Communications have increased exponentially using satellite and microwave technologies, together with digitization of data. Computers have made management of these information technologies possible on a grand scale. Even though their potential for decreased environmental impacts is greater, the actual physical component of the technology has significant environmental impacts. The average laptop uses 45 tons of raw materials and is rarely recycled. Many of these technologies use energy even when they are not in use prompting the name, "energy vampires." These technologies have also been used for business and commercial purposes. Capital is readily transferable using these communications technologies. Producers now sell to a variety of buyers that were not reachable in the ordinary sales transaction without communications technology and computers. This has increased demand for products. In the production of new products, raw materials can be purchased and transported for assembly and resale globally. The availability of transportation has also made labor itself a global commodity, with businesses moving toward the least costs of production including cheap labor. Traditional governmental restrictions on the movement of goods between marketplaces have been removed through trade agreements. In these ways, the marketplace for goods and services has been greatly expanded. The resultant growth in markets and transnational corporations has dwarfed many national economies.

If businesses thus linked into a global economy are not producing in a sustainable way, globalization threatens to intensify pressures on ecosystems, resources, and energy in ways that spiral ever more quickly toward ecosystem collapses. Meanwhile, the social and environmental dangers of globalization may be masked by the phenomenon of externalization of costs until the harms are disastrous. The ability to harness the power of markets to develop without triggering collapse of critical ecosystem functions is another area of intense controversy, even though increasing costs of transportation of goods may undercut the salience of global markets.

Globalization refers to the integration of economic, political, and cultural system across the globe. It is many things to many people. There are controversial ideas about whether globalization is a positive force that includes economic growth, prosperity, and ideas about democratic freedom or change that includes environmental devastation, exploitation of the developing world, and suppression of human rights.

The globe and its inhabitants are becoming increasingly interdependent and connected to each other as a result of technological advances in communication and transportation. National borders are increasingly porous, allowing diverse groups of people to interact. Now ideas and cultures circulate more freely.

Globalization creates new markets and potential for new wealth. It can also be the cause of widespread suffering, disorder, and unrest. Globalization has also served as a catalyst for change and social movements.

While markets have been good vehicles for increasing wealth, growth in marketplaces has also accompanied alarming trends in the rate of climate change and other environmental deterioration, as well as growth in poverty.

Globalization is a complex idea that encompasses the integration of economic, political and cultural system across the globe. The scale of contacts and ties between human societies across the world has grown exponentially, especially with the growth of the Internet. The globe and its inhabitants are becoming increasingly interdependent and connected. National borders are increasingly porous, allowing a diverse group of people to interact. This connectedness has been a technological, scientific, and now commercial phenomenon that has been spurred by technological advances in communication and transportation.

Although this sense of connectedness is not based on shared values, risk perceptions, or moral and ethical connection, it results in ideas and information circulating more freely across divides of nations, cultures, and laws. This mixing of ideas can have a positive impact; however, these changes are also uprooting old ways of life and threatening cultures that have previously been insulated from outside impact.

References

Barry, John and Robyn Eckersley. 2005. *The State and the Global Ecological Crisis.* Cambridge, MA: MIT Press.

Tolba, Mostafa K., and Iwona Rummel-Bulska. 2008. *Global Environmental Diplomacy: Negotiating Environmental Agreements for the World 1973–1992.* Cambridge, MA: MIT Press.

UNSUSTAINABLE INDUSTRIAL SECTORS

Important sectors of our economy rely on fossil fuels or other unsustainable products and processes. Some of them can change by converting to new technologies or by reengineering their products. But some of these important economic sectors will simply have to be replaced. Coal, oil, natural gas, and to some extent nuclear fuels all pose seemingly insurmountable risk of harm to the webs of life on which we depend. Yet, our global economies are dependent on them. Perhaps the greatest challenge to sustainability lies in converting these crucial sectors and their employees into a sustainable or even restorative economic function.

These sectors are currently among the most profitable in the history of the world. Companies have accumulated extraordinary capital with the assistance of tax breaks and subsidies that could be used to explore alternative, renewable energy resources. Their employees could also be reeducated and retrained for future employment in locally sustainable businesses. Some companies are devoting small amounts of capital to exploration of future enterprises in renewable energy resources. Much more of the accumulated capital is being invested in the current marketplace in things like stocks and bonds. The constructive potential of this capital is not being directly channeled into a sustainable future.

These sectors know that their economic activities contribute to climate change and other ecologically unsustainable consequences globally, yet they are making exponentially greater sums of money as this occurs. This is an example of shifting enormous unrecognized costs to future generations who will be affected in ways we do not fully understand even now. The challenge to these businesses and industry is to change, and to change quickly. The barrier to change is the amount of profit-taking that is presently going on. It remains to be seen whether investor pressures, outside pressures including government and nongovernmental advocacy organizations, and corporate ethical statements and commitments will succeed in transforming these businesses toward sustainability.

The Ceres organization, a private association of investors and environmentalists, has engaged two of these economic sectors in dialogues intended to achieve transformation toward sustainability. The Oil sector and Electric Power dialogues have begun under the auspices of Ceres and claim progress toward this goal. A similar effort at multistakeholder dialogue with some of these industries was attempted by the Common Sense Initiative.

URBAN SPRAWL

The majority of people on Earth now live in cities and towns. Urbanization allows humans to develop diverse economies that are more resistant to ups and downs. Human settlement has consequences for the ecosystems surrounding and supporting urban areas. Most human settlement is near the coasts and around rivers and lakes. Urbanization often occurs in coastal areas near fragile estuaries and wetlands. These urban areas have grown rapidly in population and geography, often without controls or services. Some of the results of urbanization include the loss of wetlands, farmlands, and forests to paving and construction.

Urban sprawl aggravates several types of environmental problems. Sprawl contributes to air pollution because people often live at a distance from where they work and public transportation is inadequate as a means of getting to work. This means there are more individual cars and trucks on our roads, contributing to poor air quality. Increased air pollution also increases other problems from climate change to human health problems like childhood asthma.

Urban sprawl also aggravates problems of equity and fairness within communities. Sprawl is associated with lack of affordable housing in desirable locations displacing established residents, and the increased development of rural land for suburban housing displacing farmers and farmlands.

Urban areas now occupy about 2 percent of the surface of Earth. The world is rapidly urbanizing with urban centers become megalopolises. There are now 19 megacities on Earth with more than 10 million residents. By 2015, some predict that 18 of the 27 predicted megacities will be in Asia. The environmental impact, or ecological footprint as it is sometimes called, of urban areas is very large. Cities and their commercial, industrial, and residential uses consume about 75 percent of Earth's natural resources. Cities take natural resources from surrounding ecosystem and biomes. They also produce large environmental impacts that can affect ecosystems outside their region. More than half the world's population of over 6 billion people live in cities. It is predicted that by 2030, there will be 2 billion new city residents. In 2030, 60 percent of the Earth's population will live in urban areas. Also in 2030, there are predicted to be 500 cities with a population of 1 million of more. This is one reason the international sustainability community focuses on the growth and development of sustainable cities. ***See also* Volume 2, Chapter 2: Land, Soil, and Forests.**

References

Benfield, F. Kaid et al. 2001. *Solving Sprawl: Models of Smart Growth in Communities across America.* Washington, DC: Island Press.

Pavel, Paloma, M. 2009. *Breakthrough Communities: Sustainability and Justice in the Next American Metropolis.* Cambridge, MA: MIT Press.

Singh, R. B. 2001. *Urban Sustainability in the Context of Global Change: Towards Promoting Healthy and Green Cities.* Enfield, NH: Science Publishers.

Brownfields: Urban Land Redevelopment

As our cities grow into megaolopolian areas with increasing populations, increasing consumption, and increasing environmental degradation of natural systems urban land redevelopment patterns shift to reclaim industrial land rather than use "greenfield." A strong policy of reusing formerly industrial land can decrease the overall environmental impacts of sprawl by increasing the density of human populations, decreasing trip generation, cleaning up threats to people and ecosystems, and preserving intact ecosystems and biomes. Brownfields is a term describing lands contaminated by industrial development. Many brownfields are now located in urban areas where industries chose to locate near crossroads or river transportation. As urban sprawl becomes recognized as a problem, the cleanup and reuse of urban lands become attractive alternatives to additional sprawl.

Cleanup of brownfields is often expensive. The costs of cleanup vary with the type of toxicity encountered at the site, the geography of the site, and the standard of clean adopted as the goal. Current owners of these sites may have to pay part or all of these costs, even if they are not the ones who made the mess. They have the right to recover their costs from the people or business that did make the mess if they can discover who they are, and if they are still able to pay anything. As a practical matter, this often means that these sites are not cleaned up. To help owners clean up these lands, the U.S. government has agreed to pay some or all of the costs of cleanup through grants and tax credits. The government also assists with brownfields redevelopment in other ways including providing insurance for these large projects.

Abandoned brownfields are often located in blighted communities. The presence of a brownfield is often a significant contribution to the blight. Another challenge involved in the cleanup and reuse of these lands is whether it should be cleaned to a standard safe for residential use, or whether it should restored to a lower standard suitable for other nonresidential use. The surrounding community may not always have a voice in this decision. A private owner of the property might make these decisions based solely on individual best interests. Political accountability for cleanup standards and reuse is greater when government funding is involved.

Sustainable Development and Brownfields

As knowledge about past and present environmental impacts on urban environments increases so will knowledge about how to develop them sustainably. There is no question that the area of cities that are toxic to people and to life generally must be contained and detoxified. These environmentally toxic areas, or brownfields, may contaminate whole ecosystems threatening other natural systems on which life depends. The extent of these contaminated areas and the rate of ecosystem degradation are currently unknown but thought to be increasing. Ideas such as environmental reparation districts are still theoretical.

A policy of community-based sustainability relates to brownfields in two ways. First, as discussed previously, it cleans up old toxic areas that present irreparable damage to systems of nature on which life depends.

It restores the environment to begin to make the ecosystem healthy. The other way brownfields relate to sustainable community development is to prevent the creation of toxic hot spots in the first place. The application of the precautionary principle to traditional land development, the creation of environmental reparation districts, and green building techniques can prevent some of the environmental impacts that led to brownfields. Part of the prevention of brownfields is the development of community-based sustainability indicators. These include measures of the environmental impacts of the human community on the air, water, and land. Transportation systems, constructions techniques, energy conservation, and industrial processes are examined closely for their potential environmental impacts and public health implications. ***See also* Volume Two, Chapter 2: Land, Soil, and Forests.**

References

Christensen, Julia. 2008. *Big Box Reuse.* Cambridge, MA: MIT Press.

Dixon, Tom et al., eds. 2007. *Sustainable Brownfields Regeneration: Livable Places from Problem Spaces.* Hoboken, NJ: Wiley-Blackwell.

Urban Ecology

Early uses of the terms *ecology* and *environmentalism* focused on wilderness lands and wild animals to the exclusion of human settlements. Cities and towns were studied by academics and regulated by politics separately and apart from nature and environment. This early disconnection is still evident in the ways in which land use and environmental laws fail to interconnect and in the ways in which environmental agencies often fail to interact with human health agencies. These disconnections are amplified in federal systems by the lack of communication among different levels of government.

As environmentalism and ecology studies have matured, they have had to reengage cities as part of the natural world and landscape. Urban ecologies are complex systems of interactions between humans and their infrastructures and natural systems, plants, and animals. Urban ecologies are also complicated by human social interactions.

To the extent that humans are viewed as a fundamental environmental problem, rather than part of an ecological system, the inclusion of human concerns in the study of environment or ecology is criticized. Some fear that issues of human social interaction like racism and urban politics will detract from the resolution of important environmental issues, or subordinate these issues to other concerns. They reject the interconnectedness of human problems and environmental issues.

WAR, PEACE, AND PEACEKEEPING

Peace and security are essential preconditions for sustainable development. War, strife, genocides, and the like create disaster for human communities that may also be visited on the ecosystems, plant and animal life, as well as destroying economies for generations to come. Given the right environmental conditions, weapons of war can create such severe damage that future generations will not have the same systems of nature that today's generations enjoy. War can leave communities unable to feed themselves or farm their lands. Paradoxically, sometimes nature—plant and animal life—can renew itself post-conflict in the absence of human life. World War II precipitated the exploration of a wide range of chemicals now commonly in use today. Before that war, these chemicals, especially those using the chlorine molecule, were not available. Now they are ubiquitous. Although war tends to drive the chemical industry's research and development agenda, it also imposes catastrophic events on ecosystems, with sometimes unanticipated results.

Certain economic measures count as domestic income spent on war-making. Wars cost lives and enormous amounts of money to prosecute. Wars can bankrupt even rich countries that throw their national income and resources into them. Nevertheless, certain businesses will make money on building and keeping the machinery of war. Sometimes they are called the "merchants of death." Many of these businesses thrive on the use of nonrenewable energy sources such as gas and oil. Environmental impacts from war business include building military installations; supplying armed forces with food, shelter, and combat armor; and waste disposal. There are several international agreements restricting environmental warfare. The 1977 Protocol on International Armed Conflicts prohibits means of warfare that may be expected to cause widespread, long-term, and severe damage to the natural environment.

Peacekeeping employs armed forces that are deployed in an effort to diminish violence and construct the basis for peace. Peacekeeping measures include preventing violence between former enemies. Peacekeeping through the development of peace parks is also thought to be another way to avoid war and multijurisdictional conflicts.

References

Ali, Saleem H. 2007. *Peace Parks: Conservation and Conflict Resolution.* Cambridge, MA: MIT Press.

Collin, Robert. W. 2008. *Battleground: Environment.* Westport, CT: Greenwood Press.

Ehrlich, Anne H., and John W. Birks, eds. 1990. *Hidden Dangers: Environmental Costs of Preparing for War.* San Francisco: Sierra Books.

UN Treaty Series No. 17512, articles 35.3 and 55.1.

WATER

Our planet is covered by water. Water is everywhere, but 97.5 percent of our planet is covered in saltwater. Humans can only metabolize freshwater, unlike some other animals, like seagulls, that possess physical tissues that filter salt from water. Freshwater occurs in obvious places such as lakes and rivers, as well as less obvious places such as underground lakes and rivers, streams and springs, known as ground water. There are even freshwater springs surging into our oceans that indigenous people identified by their difference in temperature, and that modern technology seeks to exploit.

Freshwater is essential to healthy human life. We require it for our physical integrity. Lack of clean freshwater will kill humans within a few days. Lack of sanitary freshwater is responsible for hosts of physical diseases that take a devastating toll in children.

Freshwater resources are not equally distributed throughout the planet. For humans to have access to freshwater, we must develop the mechanical infrastructure to deliver freshwater where and when we need it. Civilizations have developed such capability in different ways. The Romans developed aqueducts delivering freshwater to remote parts of their empires. The remnants of these aqueducts still exist. The ancient Minoan civilization developed indoor plumbing technology.

To provide access to clean freshwater for human populations, additional infrastructure is required to deliver and sanitize these resources. The challenge to contemporary governments is to provide funding for the infrastructure necessary to delivery and sanitize safe drinking and bathing water.

Many developed countries used tax dollars or other government funding to fund public works that build the necessary infrastructure to deliver freshwater and to clean it. Publicly owned treatment works were created to treat and clean water to safe drinking standards. These important developments were publicly funded. In other countries, especially developing countries, the necessary infrastructure for delivery and treatment may be privately funded. This means that private companies charge for the delivery of safe freshwater, rather than making clean freshwater available as a right to everyone. One consequence of this method of development of freshwater resources is that poor people cannot afford access to them. If freshwater becomes too expensive for poor people to afford, public health hazards such as cholera and other waterborne diseases may ravage a community and its economy. Epidemics of waterborne disease in developing countries can be traced to the failure of freshwater systems that had ceased to function properly in economic crises.

Clean freshwater from a tap is still a precious commodity in most of the world. In order for water to arrive ready –to use in such a state, many processes must be in place. Infrastructure transporting water from its source in nature must be constructed. Sanitation processes are probably required to be applied before it arrives at a home tap. These sanitation processes also require substantial investments in infrastructure, maintenance, and services to keep freshwater resources safe and potable.

Sometimes government itself will build and maintain all of these required structures to provide reliable, sufficient freshwater to its people. In these countries, publicly owned water companies are often referred to as publicly owned treatment works. These are a type of publicly owned utility service. Alternatively, some governments allow private business to own and control some or the entire network supplying clean fresh water to paying customers. Privately owned water systems may be more attractive to development banks and investors, and thus easier to build. The cost of developing and delivering clean fresh water probably will escalate as climate change and other ecological events increase the acidification and salinization of available freshwater.

What happens to people who cannot afford to pay the price of clean freshwater? Moreover, why should business care about those consequences? Water for drinking and cleaning our bodies and our food is a necessity of life. Without adequate supplies we die, and children die much faster from dehydration than adults as a result of their smaller body size. Poor people who cannot afford to pay the price for clean water will suffer the consequences in terms of higher death and morbidity rates among adults and children. They will also be exposed to greater waterborne diseases driven to unclean water sources to the extent they are available. In densely populated communities, waterborne health hazards will travel rapidly throughout the population, including those who have not had direct contact with polluted water. Class is no exception. Poor workers may infect those with whom they work, including domestic workers. Water sources may become contaminated, resulting in higher costs. In the 19th century, these consequences led the leaders of the Industrial Revolution to institute public health processes in sanitation that led to the creation of public water treatment and other sanitary measures. ***See also* Volume 2, Chapter 2: Water Systems.**

References

Griffin, Ronald C. 2006. *Water Resource Economics: The Analysis of Scarcity, Policies, and Projects.* Cambridge, MA: MIT Press.

Hlavinek, Petr et al. 2009. *Risk Management of Water Supply and Sanitation Systems.* New York: Springer.

Just, Richard E., and Sinaia Netanyahu. 1998. *Conflict and Cooperation on Trans-Boundary Water Resources.* New York: Springer.

Troesken, Werner. 2008. *The Great Lead Water Pipe Disaster.* Cambridge, MA: MIT Press.

CHAPTER 5

Future Directions and Emerging Trends

Developments in terms of sustainability are happening on a daily basis in many different areas. The daily news from multiple sources informs us of new scientific predictions, developments in technology, losses of species, ecosystem damage and failures, human creativity, and community responses. Some of these indicators are profoundly saddening, and to some unexpected. Other indicators are sources of hope and optimism.

Public and private voices now clamor for sustainability as the environmental news on the limitations of Earth's natural systems becomes observable and creditable. Many policymakers can no longer escape into easy decisions that reaffirm traditional methods of public policy analysis. Reaffirming past decisions that negatively affect the environment only add to the accumulation of environmentally degrading activities. Yet, there is no clear path or direction for these policymakers. The old reliable parameter of sound public policy, science, no longer provides enough information fast enough. Old methods of proof of causality may work in the laboratory but not in the public arena. Public policy for sustainable development is fraught with great uncertainty and risk. Post-normal science, ecological risk assessment, application of the precautionary principle, and thorough and accurate environmental monitoring may decrease some of the uncertainty and risks but may uncover others.

One public policy approach is to look for trends in decision making around sustainability and for indicators of good sustainable policies. As our approach to sustainability becomes globalized, the approaches to sustainability greatly increase. Many places in the world are trying different methods and approaches to be sustainable. They are doing so with international aid and assistance and in communication with global networks. Some of these places have very similar values, others have very different values, and others have not examined their core values. Most of these places do share one concern however; they want to develop policy that is sustainable and within the carrying capacity of their ecosystem. ***See also* Volume 2, Chapter 2: The Language of Sustainable Business.**

Reference

Heal, Geoffrey. 2000. *Valuing the Future: Economic Theory and Sustainability.* New York: Columbia University Press.

CHANGING OUR THINKING: FROM INEXHAUSTIBLE TO IRREPLACEABLE

One of the most effective ways to achieve change is to change the story that we tell ourselves when we are trying to make sense of complexity. It is human nature to see and think what we already know and what we want. A problem like sustainable development may be more complex than anything in our recorded history. The level of uncertainty overwhelms our most trusted leaders, our best scholars, and our brightest diplomats. It is a dynamic complexity that resists simplification and static description. As cultures try to make sense of the complexity of sustainability, the stories told must also change, which require changes in traditions and deeply held values.

Religious myths of natural superabundance and infinite providence have a powerful hold on the ability of humans to imagine our relationship with the Earth. These creation myths also tend to support the idea of nature's instrumental value to humans, a human-centered approach focused on the utility of nature to humans.

The ability to change the story of our human relationship with our planet from one of righteously privileged user toward an appreciation of fragile and irreplaceable nature of the dynamics that support life as we know it is powerful. The power to revise these deeply held intuitive views enlists many other change methods without requiring explicit agreement or instructions. Changing such perspectives requires equally compelling cultural views. Contemporary media from films and television to the Internet have made a powerful counter-message of the fragility of the webs of life on which all living beings depend. Religions have joined in the Earth Charter to try to bridge these conflicting stories and provide the basis for transcending the barriers that confront us in our journey toward sustainability.

CLEAN PRODUCTION TECHNOLOGIES

Sustainable production methods will require a shift in values that penalizes waste and pollution, as well as incentives reductions in wasteful consumption. Domestic policies on the lowest levels of government can help if they emphasize and facilitate nonwasteful and nonpolluting production methods, including efficient use of energy.

Nonfossil Fuel Economy: Hydrogen Economy or Carbohydrate Economy

One of the challenges of contemporary life is to imagine what a differently configured economy might be like. That is not at all clear because we are dependent on fossil fuels for the energy and products that define contemporary life. Some have forecast that our economy could package and use hydrogen much as we use fossil fuels now if the technology for capturing hydrogen in fuel cells can be perfected. Others have forecast a similar future for an economy based on energy from the burning of carbohydrate sources. Experimentation with the carbohydrate model shows limitations based on failure to consider the full lifecycle of cultivating and producing this energy source. The true and full cost of these carbohydrate fuels like corn-based ethanol appears to be greater than their fossil-fuel counterparts. In addition, disruption in the food supplies for many poor countries and their people is proving significant, and there is no transitional plan to deal with disruptions from this source.

The search for one technology or product to completely (and inexpensively) substitute for the technology in which we have presently sunken our investments and hopes, however, is a little like a search for a legendary silver bullet. The observable truth is that we have many options for alternatives that allow many different accommodations to local conditions and adaptations of scale. Figuring out a coherent way of patterning a variety of options is a more comprehensible strategy, if only for a transitional generation or two.

An example of this dynamic is the growth and development of green nonprofit centers. Some of these centers offer financial investors sustainability-based value expression by supporting many other nonprofit environmental organizations. The Thoreau Center for Sustainability in New York City is an example of a sustainable and green nonprofit center with 12 nonprofit organizations sharing office space. These centers are examples of alternative pathways to reach sustainability.

Reference

Hess, David J. 2007. *Alternative Pathways in Science and Industry: Activism, Innovation, and the Environment in an Era of Globalization.* Cambridge, MA: MIT Press.

COLLABORATIVE DECISION-MAKING PROCESSES

Collaborative environmental decision-making processes involve multiple stakeholders from the community, government, and industry to work together to solve an environmental problem. This is opposed to adversarial environmental decision-making methods that use courts and judicial processes to determine a winner and a loser. Other issues regarding the limitations of judicial approaches are that access to courts is restricted by income, politics, and legislation. This is especially true in environmental conflicts. There are few if any laws about sustainability, but there is a global consensus that sustainability is a major social goal. By restricting judicial accessibility, this forum makes incomplete and transient decisions that may actually impair sustainability. As environmental impacts accumulate and conflicts increase, new collaborative approaches are developing.

Collaboration in the use of natural resources offers benefits to sustainable approaches. It is being used more in watershed decisions and in some complex urban environmental issues. Collaboration allows for an evaluation of the full range of ecosystem services because of the multiple sources of information. It can increase the range of value perspectives on that use and the range of options for protecting and conserving that natural resource. By increasing information, value perspectives, and range of options, collaborative environmental approaches can also decrease uncertainty because the implications of various options are better known. In this way, uncertainty about the application of the precautionary principle to environmental decision making is decreased.

Collaborative environmental approaches may also be a way to span political boundaries that are not based on natural features. Because of this, collaboration may be confrontational as stakeholders from other nations or cities are included. Collaborative environmental approaches may also become controversial, as formerly excluded groups are brought in as stakeholders. Eroded environmental resources, such as water, can present public health threats to formerly excluded groups. Issues of reparations and compensation for these damages may arise. This adds to the financial costs of collaborative environmental approaches, which are generally more expensive than traditional ways of environmental decision making. Collaborative environmental decisions may cost more because of the time and money of inclusion of new stakeholders and the need for capacity building in that system for complete and accurate ecosystem knowledge. Sustainability advocates, however, point out that those collaborative approaches may save money in the long run because they provide a fuller accounting of ecosystem services.

Collaborative approaches for sustainable environmental decision making also require that stakeholders are interested in the common good, not just self-interest. Cooperative, forthcoming, and engaging dialogues work better than adversarial, interest group, conflict-based negotiation in collaborative models. Alternative dispute resolution is not functional in collaborative sustainability decision making. The facilitation of multiple stakeholders, especially those with different capacities, is usually needed in the early stages of collaboration. The facilitator makes sure that all stakeholders have all the information, that they understand the information, and that they understand each other.

Collaborative approaches can be profitable for businesses that locate or create market niches, or spaces where demand for their product is expressed. For example, with the rise of green, plant-based roofs, the demand for functional plants increased. Although nursery's have always tried to meet local demand with local and exotic plants, the market for ecologically functional plants is just starting. Plants can perform many functions, including cleaning wastes from water, retaining necessary water, and ecosystem restoration. Cattails and bulrushes are used to treat wastewater wetlands because they lower levels of nitrogen and phosphorus, common agricultural chemical runoff from fertilizers. Their roots gather up toxins and help break them down. Some plants are used in a waste cleanup process called phytoremediation. In this process a plant, usually a tree with a deep tap root such as a willow or poplar, is planted in a contaminated area. It grows rapidly, gathering (sequestering) toxic chemicals such as cadmium and breaking them down.

Collaborative approaches between local nurseries, environmental agencies, and communities is one approach to sustainability under a market model.

References

Baber, Walter F., and Robert V. Bartlett. 2005. *Deliberative Environmental Politics: Democracy and Ecological Rationality.* Cambridge, MA: MIT Press.

Sabatier, Paul et al., eds. 2005. *Swimming Upstream: Collaborative Approaches to Watershed Management.* Cambridge, MA: MIT Press.

Stern, Alissa J. 2000. *The Process of Business/Environmental Collaborations: Partnering for Sustainability.* Westport, CT: Greenwood Publishing.

EDUCATION ON THE ENVIRONMENT FOR SUSTAINABILITY

Educational institutions have been engaged in planning and thinking about sustainability for at least two decades. Most of their efforts have been aimed at achieving sustainability in their operations much like any other business organization. These efforts include lowering environmental impacts from their physical plants through reductions in toxins and waste in areas including grounds, housekeeping, and food services. In addition, they have focused on achieving energy efficiency in the construction and renovation of buildings, and the use of energy for light and heat. Some also encourage these types of measures in their suppliers and contractors. Some have also tried to incorporate long-term vision for sustainability in their financial operations including endowment investments and supply chain management. Many university sustainability plans engage in transportation by increasing public, pedestrian, and bicycle transit. Many universities are developing an "ecology" house, which can be a residential community, and is often designed to be sustainable. Some universities are protecting and reintroducing wildlife on campus. Composting of wastes, urban farming, community gardens, and sustainable landscaping are all increasing on college campuses as universities meet student demand for learning about sustainability.

Curricular commitments to sustainability as a teachable subject remain more elusive. One reason for this curricular disconnection may lie in the way that higher education has organized its disciplinary structure in contemporary colleges and universities. The modern university is organized around distinctions between the arts, sciences, and professions. Liberal arts are usually based on the classical areas of learning including history, philosophy, rhetoric and politics, and mathematics. Sciences are often distinguished by the branches of scientific learning that developed during the enlightenment period, including biology, chemistry, geology, and physics. These distinctions are the basis for the degree-granting and teaching structures within most universities. Sometimes, these distinctions are compared to agricultural silos in the sense that they keep various fields separate and coherent, but they prevent mixing that might result in new combinations.

Sustainability crosses the frontiers of multiple disciplines, from social sciences, to physical sciences, and into areas like law, business, and engineering, which are often associated with professional training.

References

Allen-Gil, Susan et al. 2008. *Addressing Global Environmental Security through Innovative Educational Curricula*. New York: Springer.

Leal Filho, Walter, ed. 2005. *Handbook of Sustainability Research: Environmental Education, Communication and Sustainability.* New York: Peter Land.

Rappaport, Ann, and Sarah Hammond Creighton. 2007. *Degrees That Matter: Climate Change and the University.* Cambridge, MA: MIT Press.

Operations and Curriculum

Many institutions have moved their operational conduct toward sustainability. Campuses often operate as businesses do, including their grounds keeping, housekeeping, food services, building, and housing functions. In addition, campuses resemble businesses in their relationships with suppliers and contractors. They may also have substantial sums of money to manage in the form of endowments from benefactors. Many campuses have found that reducing their environmental footprint in these operations is attractive to their students and makes practical sense in both short- and long-term financial efficiencies. Many institutions of higher education have moved toward an emphasis on sustainability in their operations.

Moving toward sustainability in terms of what is taught in the curriculum has been significantly more difficult to achieve. This is likely due to the organization of the disciplines within the contemporary university system, as well as their systems of rewards within the different disciplines. In addition to challenging traditional differences between disciplines, sustainability as a field of study is new and embraces uncertainty in a way that is uncommon to the traditional fields of learning.

Reference

Barlett, Peggy F. and Geoffrey W. Chase. 2004. *Sustainability on Campus: Stories and Strategies for Change.* Cambridge, MA: MIT Press.

Epistemological Ignorance: The Challenge of Sustainability

Education at all levels focuses on teaching what we know and what we are reasonably sure about based on facts and inferences. At the higher and more abstract levels, education attempts to teach ways to think about what we are less sure of when facts are uncertain and inferences are attenuated. That is one level of uncertainty. When faced with uncertainty at this level, education responds with philosophies and values based on discussions that can be charted and certified by degree-granting programs. A different experience of uncertainty occurs when we do not know what we do not know. At that level of uncertainty, traditional education has had difficulty in conferring degrees and pedigrees. Many of the issues involved in the study of sustainability involve this second level of uncertainty. Beyond conflicting and competing values invoked by topics such as those noted in the controversies section of each volume, the complexity of factual data that comes from each place studied makes education a moving target, crossing many academic terrains, and occasionally requiring the honest admission of ignorance and humility.

The ultimate challenge for education in this situation is how to teach about complexity and ontological and epistemological ignorance. The ultimate challenge for a society struggling with issues of sustainability is how to make decisions under these circumstances.

Some meta-disciplines have long struggled with these levels of uncertainty and the need for decision making. One such discipline is law. Law confronts the need for certainty in facts with a procedural acknowledgment that fact is socially constructed. In this discipline, fact-finding is done by a jury of social peers recognizing that their conclusions are reflective of the community's own cultural perspectives, not necessarily a scientific enterprise. They are often instructed in how to resolve uncertainty in terms of a spectrum of probabilities ranging from a mere preponderance to a moral certainty beyond reasonable doubt.

When asked to make policy recommendations, scientists may experience the discomfort of approaching disciplinary limits in the form of uncertainty and the political consequences of uncertainty. Theories such as post-normal science attempt to place traditional sciences within a matrix of other knowledge such as community experience for the purpose of escaping the paralysis of judgment that uncertainty can impose within traditional limits.

These multidisciplinary, meta-disciplinary approaches to the problems of uncertainty are not yet comfortably housed within the architecture of contemporary educational structures, challenging educational institutions to come up with new structures.

Public Outreach

Currently, public outreach about anything sustainable is minimal. In the United States many still perceive environmental regulation as "liberal" and as an impediment to job creation and economic growth. The lack of public outreach causes a lack of public knowledge about sustainability. This in turn causes a lack of public engagement with election of candidates that promise to protect the environment, and this in turn stifles the development of research and public policy in the area of sustainability.

In some pioneering parts of higher education, however, new research is emerging about sustainable approaches in industry, government, and communities. Neighborhoods in Seattle, Washington, are starting a grassroots effort for sustainability.

The city of Seattle has formed an Office of Sustainability and Environment. The goal is to collaborate with other city agencies, nonprofit groups, business organizations, and neighborhood organizations to provide outreach to the public about actions that advance sustainability. It has emphasized environmental actions and results. The mayor wants to make Seattle the green building capital of the United States. One way to meet this goal is to reduce greenhouse gas emissions, and he has agreed to try to reduce levels to 7 percent below 1990 levels by 2012, and 80 percent of 1990 levels by 2050. He wants to increase energy efficiency in new and existing buildings by 20 percent. He has developed an extensive public outreach program to meet this goal. He has also developed other programs that save Seattle's trees, create a community greenhouse gas inventory, creates a community carbon footprint, and develop an implementation plan.

Neighborhoods are a particular focus in this example of a public outreach effort. Seattle has created a Sustainable Urban Neighborhoods Initiative. This initiative works closely with neighborhoods to evaluate their strengths and weaknesses in terms of sustainability. It organizes and facilitates meetings among neighborhoods, universities, and city agencies. It does on-the-ground research about environmental conditions in that neighborhood. It also helps increase the capacity of neighborhoods to understand what the research and indicators mean to the residents.

All across the United States, many large urban areas are grappling with the costs of ignoring centuries of environmental impacts and hoping that public outreach for sustainability will improve the economic, ecological, and equitable aspects of the quality of life.

References

Blewitt, John. 2008. *Community Development, Empowerment and Sustainable Development.* Devon, UK: Green Books.

Gross, Julian. 2005. *Community Benefits Agreements: Making Development Projects Accountable.* San Francisco: Good Jobs First.

Lewis, Sanford J. 1993. *The Good Neighbor Handbook: A Community-Based Strategy for Sustainable Industry*, Waverly, MA: Good Neighbor Project.

EQUITY

Businesses that have taken short-term transitional losses to eliminate waste and toxins in their products and production methods have been rewarded in long-term economic gains, as well as creating measurable improvements in their environmental impacts. These businesses have embraced the linkage between environment and economy, but they have not necessarily incorporated communities and their well-being into this new model.

GLOBALIZATION AND LOCALIZATION

Transportation of goods limited because of costs of energy for transport increasing dependence on local producers, requiring more diversity of local production of goods. As transportation costs increase, the shipments of diverse goods from distance places decreases. Transportation is also global and much quicker than ever before, but the dependence of transportation companies on oil and gas has made this network unsustainable. The assumptions around the cost of transportation and the ability to transport goods may affect the type of globalization we see in the future. Globalization of communications facilitated globalization of markets and capital. Globalized markets for goods depend on a transportation network that is dependent on fossil fuels. As these fuels become more expensive, and their supply less reliable, the reliance of globalized trade on these transportation networks will change, sometimes dramatically. When the cost of transportation of goods becomes too expensive for consumers to bear, they will be forced back into localized goods and products. This is a counter-measure to the effects of globalization moving employment in certain manufacturing and agricultural sectors to distant locations in order to save labor costs. When transportation costs exceed labor costs as a cost of production, localized supplies will regain their market saliency.

Agriculture in particular will need to adapt to the high cost of transportation. Some foresee that this will force localization of foods, meaning increasing the amount of agricultural production near urban areas. The ability to find usable land in urban areas has been severely compromised by urban sprawl and the environmental impacts of population growth in urban centers, taking both land and water resources away from local agriculture. Even in many areas of rural counties where land was set aside for agricultural preservation, rapid suburban development characterized by sprawl continues to compromise the ecological ability of that land to support the population.

One idea for localization of food resources in urban areas is vertical farming. Traditional farming is done on the surface of the earth at the level of the topsoil. Among the challenges of traditionally configured farming is the need to acquire and manage large tracts of land usually outside of the market area for which the food is grown. Acquiring large tracts of arable land is expensive and may destroy wetlands or impinge species habitat. In addition, managing such tracts may require substantial investment in machinery and infrastructure such as irrigation or drainage, and farm equipment that uses fossil fuel for its energy. Vertical farming is more similar to greenhouse growing operations, done in multistory buildings resembling skyscrapers. It is compact in its impact on the actual surface acreage, preserving wetlands and other important land uses including species habitat. In addition, locating these sky farm buildings in and near urban markets will greatly reduce the cost of transportation as part of the cost of food. ***See also* Volume 2, Chapter 2: Globalization; Volume 1, Chapter 4: Agriculture.**

INFORMATION

Information about ecological and environmental conditions, as well as human health and economic information will be critical to sustainable decision making in the future. Currently, basic data are missing, important information is not shared, and information sources are ignored or corrupted. Communities are invaluable repositories of information about the history and human uses of land and resources. Community knowledge must be systematically collected and considered in sustainable decision making. Scientific knowledge has been significantly distorted or discarded by some governments. This corruption of data impedes sustainable public policymaking.

Information about environmental and ecological conditions is collected by various levels of government that do not share it with each other. Creating forums in which such information is shared and processes for sharing it is important to integrated decision making. Land use and developmental decisions should consider human health and environmental data.

SHARES: LOAD SHARING, CATCH SHARING

Sharing as a way of doing business may seem anticompetitive at first glance. A competitor helping another competitor, or sharing scarce resources with a competitor, might seem to be counterintuitive. Sharing in order to lessen environmental impacts, however, is a promising technique in several areas.

Load Sharing

Trucking is a critical link in the transportation of goods from their sources to the ultimate consumer or end-user. The price of fuel has severely impacted truckers and their companies. In an effort to save fuel, reduce transporting empty space, diminish traffic jamming from trucks, and reduce carbon impacts, several companies are experimenting with the idea of sharing truck space. The way this concept works is for companies, sometimes-fierce rivals, to share space in trucks so that these trucks will not be driven empty. If a delivery is made to one center, rather than returning empty to another location, the truck may take on the goods of a rival. This requires sharing of delivery schedules and information with a dispatching function. This dispatch function may be provided by government or through mutual negotiations.

Catch Sharing

Fisheries are finally becoming recognized as fragile, exhaustible resources, much like a commons in the Tragedy of the Commons. After some famous fisheries collapsed, leaving their communities without an economic base, human communities are embracing the need for proactive fishery management. Some scientists have warned that there will be virtually no wild seafood by the middle of this century unless immediate actions are taken to protect ocean fisheries from overfishing and collapse.

One solution to manage these limited resources in ways that balance the need for economic security with the ecological limits of the species and area is to divide the total catch into shares and distribute these shares among the fishers. These programs are called Limited Access Privilege Programs (LAPPs). The claimed benefits of these programs are to provide fishers with a stake in the continued health of their fishery, and an incentive to avoid waste, a serious problem in fishing called "by-catch." One criticism is that these programs effectively privatize a public resource. Project design may determine the ultimate ownership of these shares and their worth to communities dependent on these resources. *See also* **Volume 2, Chapter 4: Tragedy of the Commons.**

Reference

Webster, D. G. 2008. *Adaptive Governance: The Dynamics of Atlantic Fisheries Management.* Cambridge, MA: MIT Press.

SUSTAINABILITY ASSESSMENTS AND AUDITS

Although the environmental assessment process offers the best foundation for a sustainability assessment, its application to policies and practices increases under sustainability.

Most institutions and organizations have policies and practices that contribute to environmental impacts, sometimes in an unsustainable manner. Although environmental audits were developed to help industry comply with environmental laws, they did not reach broad application to nonindustrial sectors such as residential or institutional uses like schools. For an audit to be useful as a sustainability audit, and for it to be a building block of a sustainability policy, it must apply to all possible human impacts because all of them can contribute to increasing cumulative impacts. Sustainability audits are in a dynamic state of growth. Currently little financing or law requires them, but they are the tool that will implement the sustainability missions in international documents and agreements.

Most institutional sustainability audits cover all environmental impacts from the organization. Organizations can differ greatly in size, function, mission, and environmental practices. Churches, municipal offices, schools, state offices, and neighborhood organizations are a few examples of nonindustrial institutions and organizations. Many of these organizations are interested in being sustainable, or at least learning more about it. Some may not be able to afford a "sustainability audit" or fear that it may reveal environmental liabilities. Some environmental advocacy groups are promoting the idea, and some consulting groups have begun that expand from the usual corporate practice into municipal and state clients.

The primary focus a generic sustainability assessment of an organization is energy usage, waste production, water usage, paper usage, staff and human resource policies, travel practices, and purchasing practices. The ultimate goal is to know the organization's ecological footprint. Ecological footprint analysis measures the amount of renewable and nonrenewable ecologically productive land area required to support the resource demands and absorb the wastes of the organization.

In terms of energy, the goal of a sustainability assessment is to reduce consumption, and certainly any wasteful or inefficient practices, which can vary greatly depending on the climate. Energy conservation is often a sustainable practice that saves money for most consumers. Energy should also be procured from renewable sources, even if they are more expensive in terms of cash cost. This may require some research and investigation because some renewable sources can be on site. Solar panels, for instance, could be added to structures housing the organization. Sustainability assessments also emphasize the use of energy-efficient equipment and energy infrastructure. Sensor-activated appliances such as lights, escalators, and people movers are prioritized. The use of passive systems such as awnings and use of cross-ventilation from open windows is also assessed. Designs and practices that use natural light as much as possible are also part of an audit.

In terms of waste overall, the sustainability audit focus is to reduce use, and to reuse and recycle all wastes. Waste issues are important in a sustainability audit. In some industrial ecology approaches to sustainability, the idea is to close the loops of waste production. This often requires a search for markets for what was once disposed of as waste. In many organizations, paper waste is an issue because the use of computers and printers increased demand for paper. Many offices now try to go as paperless as possible to reduce paper waste. Two-sided copying and printing are encouraged in organizations with a sustainability ethos. Paper recycling is highly encouraged and is made as convenient as possible. The type of paper can be important. Using paper for printing, paper towels, and toilet paper from recycled or sustainable sources is also assessed. In terms of other wastes, as much of it as possible should be recycled, including toner cartridges, aluminum, glass, plastics, and packaging. Obsolete objects of technology, such as older computers, should be recycled or donated if possible. Food implements should be reusable, and organic waste should be composted.

The environmental impact of the institution on water is important in a sustainability audit, especially if water is a threatened natural resource in that location. Here a sustainability audit would seek to minimize water use and to pretreat it before it enters natural water systems. Some ways to reduce water usage is to use motion-sensitive faucets. Motion sensors can also be used on hand air dryers to save energy and paper. The use of toilets that provide two different types of flushes—one for solid waste and the other liquid waste—is another way to reduce water usage. Sustainability audits also examine the land around the organization in terms of landscaping practices. Generally, the use of indigenous species requires less fresh water than the traditional grass lawn. Environmentally sensitive irrigation and pest management programs can also lessen environmental impacts on natural water systems. The choice of cleaning materials and soaps is important because many of these are simply washed down the drain. The use of nonphosphorous soaps is another way of reducing impacts. Impacts on water resources can also be reduced by using gray water. This is water directly from the roof or other catchment area, and water sinks and drains from cleaning and bathing. Gray water can be used for landscape water maintenance, toilets, and external washing. Knowledge of the complete chemical content and biodegradability of all indoor and outdoor cleaning materials is essential for accurate knowledge about the ecological footprint of the organization. Many organizations in the United States file a Material Data Safety Sheet with the Occupational Safety and Health Administration (OSHA).

A sustainability audit useful for generic application to most institutions and organizations also examines the organizational process and people. There are usually three areas of emphasis: staff and human resources, transportation, and procurement.

In terms of staff and human resources, the sustainability audit will assess the understanding and awareness of issues that the particular assessment views as relevant to sustainability in that organization. Because sustainability views can range from sustainable quarterly profits for a corporation to radical swings in values and consumptive behaviors by communities, the assumption of the sustainability assessment should be stated. An assessment could examine efficiency in production as part of a sustainability audit. Generally, a generic sustainability audit will look for environmental information points and forums, communication around environmental issues like procurement and recycling, specific positions such as a sustainability coordinator, and other social activities. Continuing education programs in professions and trades that encourage staff to learn about sustainability is also part of this aspect of the sustainability audit.

Transportation and travel of staff can dramatically increase the carbon use of an organization. Most alternatives to private vehicles are encouraged, such as walking, bicycling, and mass transit. The provision of storage for alternative transportation and showers for cyclists is also assessed. Telecommuting, or working at home via the Internet and other communication systems, is also encouraged in the audit. In organizations with large fleets of vehicles, the sustainability audit will examine car pools, purchasing of carbon dioxide emission offsets, and use of fuel-efficient and low-emission vehicles.

Most institutions and organizations have to purchase goods and procure services. In terms of a sustainability audit, a lifecycle impact assessment for all purchased goods is evaluated. This means that how the product is manufactured, used, and disposed of is included in the computation of ecological footprint. Recycled paper and other products are highly prioritized. In examining the supply chain, the purchaser should support local suppliers if transportation costs of distant suppliers increase environmental impact. Purchasing supplies in bulk can also lower transportation impacts on the environment by decreasing the number of delivery trips. In terms of manufacturing industries most turn to on-shore manufacturing sites to reduce their impact. Whether this reduces their carbon footprint, is still a matter of debate. In terms of office supplies, the use of low volatile organic chemical emissions from paint, carpets, glues, and sealants is preferred under a sustainability assessment. What consumers see most is retail trade businesses.

One of the leaders in sustainability is IKEA, a large multinational corporation that sells house wares and furniture. IKEA operates about 285 stores in 36 countries with sales of about $27.9 billion to 500 million customers. About half of its carbon footprint comes from transportation of goods. To decrease costs and environmental impacts, IKEA builds new stores near ports or railroad depots to avoid truck shipping. It has a 2006–2009 internal sustainability plan that focuses on supply chain management. It requires an amount of responsible forestry, green building, and reductions in green house gas emissions from all its suppliers. IKEA is very effective at eliminating waste in all areas, especially energy usage.

In the United States, IKEA's goal was to reclaim 90 percent of the store waste, and latest figures from their annual report show that they are reclaiming more than 84 percent of past waste. IKEA's case may be different from other retailers because it emphasizes affordable prices and takes strong measures to achieve them. If it can decrease costs by shortening its supply chains, then it does so. This is sometimes referred to as "eliminating the middleman." If shortening the supply chain also meets its sustainability plan goals, then so much the better from IKEA's perspective. IKEA leadership in sustainable business practices is being implemented in whole or in part by some other big box retailers such as Wal-Mart. Sustainability assessments are becoming important tools to engage in local and global business.

A generic sustainability assessment for noncorporate institutions and organizations is a large, dynamic, and necessary step for sustainable

International Organization for Standardization

Another set of standards internationally used by industry is being developed by the International Organization for Standardization (ISO). The ISO is the world's largest developer and publisher of International Standards. It has 157 member countries and is itself a nongovernmental organization. It developed a standard used in environmental audits called the ISO 14001, although it was not developed specifically for sustainability. Most ISO products are developed for a specific product or process. ISO 14001 designs a process-based approach to environmental management of an industrial plant. It does have the ability to monitor a wide range of environmental issues, and it offers a potential platform for a sustainability assessment. The ISO 14001 process can be applied generically. It does not list any performance goals or criteria and relies on internal audits.

policy development. Corporate institutions, college campuses, National Parks, and communities all do their own sustainability assessment. They all have different purposes and functions because they are used by different stakeholders.

Sustainable Assessments of Trade Policies

The strong interest in sustainability in the international community combined with a global concern about all ecosystems pushes sustainability assessments to trade policies. One of the environmental concerns is that as global markets open and expand, some natural resources will be irreparably damaged. Some international environmental organizations advocate for the use of sustainability assessments to shape trade and investment agreements so that they support sustainable development. Some of these types of trade sustainability assessments go beyond environmental measures and include social and equity measures.

Various industries use different standards to measure these parameters. One that they prefer is the Global Reporting Initiative (GRI). The GRI is a large multistakeholder network of experts around the world who participate in GRI, use the GRI Guidelines to report, access information in GRI-based reports, or contribute to develop the Reporting Framework. ***See also* Volume 2, Chapter 3: United States.**

References

Fort Carson Sustainability and Environmental Management System Plan, www.fedcenter.gov/_kd/Items/actions.cfm?action=Show&item_id=10217&destination=ShowItem.

Gibson, Robert B., et al. *Sustainability Assessment: Criteria and Processes.* London, UK: Earthscan.

Global Reporting Initiative. www.globalreporting.org/AboutGRI/WhoWeAre/.

Helming, Katharina et al. 2008. *Sustainability Impact of Land Use Changes.* New York: Springer.

Kassim, Tarek A., and Kenneth J. Williamson. 2005. *Environmental Impact Assessment of Recycled Wastes on Surface and Ground Waters.* New York: Springer.

SUSTAINABILITY INDICES

There are many different indices published by different groups relevant to the task of assessing sustainability. Some focus on environment and ecosystem status, some focus on social indications and human health, others focus on economic indicators, and still others focus on government. Aggregators of data will combine various measurements to analyze sustainability for their purposes. In considering the value of these types of indices, it is important to know what they include, and what they do not include.

Financial data aggregators have sustainability indices that are different from governmental corruption aggregators. Some of these indices rank organizations, countries, or local governments. Their methodology may differ widely in terms of the choice of relevant data, and the choice of comparisons or rankings between similar organizations. For example, the Environmental Performance Index (formerly the Environmental Sustainability Index) ranks 149 nations based on 25 different indicators of sustainability. It does not include in this ranking social or economic justice data but focuses instead on environmental indicators. Other indexing organizations will include in their rankings information about governmental corruption, equality of income distribution opportunities for women, and health data about children. These additional factors are important to predictions about stability and risk of investing in certain areas. Still other organizations rank corporations and their internal commitments to sustainability. These rankings may incorporate data including relationships with employees, suppliers, and contractors.

References

Breitmeier, Helmut, Oran R. Young, and Michael Zurn. 2006. *Analyzing International Environmental Regimes: From Case Study to Database*. Cambridge, MA: MIT Press.

Felleman, John. 1997. *Deep Information: The Role of Information Policy in Environmental Sustainability*. Westport, CT: Greenwood Press.

Energy Use

Industry is the largest user of nonrenewable energy. If industry and businesses achieve ways to either conserve energy, or decrease the intensity of nonrenewable energy use in their businesses and industries, a substantial step toward energy sustainability would be achieved. Energy intensity is defined as the units of energy required to produce an item or measure of production like gross domestic product. Companies can change their behaviors toward greater sustainability by both achieving greater efficiency in terms of their productivity with the energy they do use (producing more with less energy), and also by conservation measures that include changing to renewable sources, as well as reducing or eliminating use of nonrenewable energy sources. Companies striving for increased accountability should report data about both their conservation measures and their intensity of energy use.

Reference

Hoffman, Peter. 2002. *Tomorrow's Energy: Hydrogen, Fuel Cells, and the Prospects for a Cleaner Planet.* Cambridge, MA: MIT Press.

Greenhouse Gas Emissions

Climate change and greenhouse gas emissions are among the most critical items to identify in assessing the sustainability of business practices; however, there are many different ways to quantify and report data relevant to this item. The Greenhouse Gas Protocol, a joint project of the World Resources Institute and the World Business Council for Sustainable Development, has become the international standard used by many other reporting agencies for determining the greenhouse gas emissions of an organization. This accounting process relies on two standards linked together. The first is a calculation tool called the corporate standard that is meant for use by both private and public organizations. These standards are further refined by industrial sectors. Second, there is a protocol for offsetting the effects of mitigation and conservation activities. This information is often incorporated into other aggregated lists ranking organizations in terms of sustainability or environmental achievements.

References

The Greenhouse Gas Protocol Initiative. www.ghgprotocol.org/.

McIntosh, Roderick J., Joseph A. Tainter, and Susan Keech McIntosh. 2000. *The Way the Wind Blows: Climate Change, History, and Human Action.* New York: Columbia University Press.

Yang, Zil. 2008. *Strategic Bargaining and Cooperation in Greenhouse Gas Mitigations: An Integrated Assessment Modeling Approach.* Cambridge, MA: MIT Press.

Waste Sent to Landfills

Much of the waste filling up municipal landfills called municipal solid waste comes from industry, especially construction materials that could be recycled or repurposed. Companies that are trying to achieve clean and sustainable production often achieve significant gains by reducing this material throughput that would otherwise be waste. The cost of waste is high and falls not just on the company in the form of the cost of waste transfer and disposal, but also on communities in the form of municipal costs associated with providing waste repositories or landfills, and on communities that must live in close proximity to these depots for waste.

Companies can easily identify and report the amount of their operation that ends up in a landfill or other solid waste repository.

References

Princen, Thomas, Michael Maniates, and Ken Conca, eds. 2002. *Confronting Consumption.* Cambridge, MA: MIT Press.

Rogers, Heather. 2006. *Gone Tomorrow: The Hidden Life of Garbage.* New York: New Press.

Materials Balance: Inventory and Throughput

Environmental auditing and inspecting have generally been done by attempting to measure emissions. Emissions monitoring may not capture all emissions. Emissions can be affected by heat and other chemicals in the processes of manufacture and change media from solid to liquid or gas state. A different method of calculating emissions from a business facility, or process of manufacture, is the mass balance or materials balance method. Rather than measuring outputs or emissions, this method determines output or throughput from input or the actual materials used.

In achieving sustainability, a company that reduces waste as throughput and uses its materials without waste is achieving an important measure of sustainability. This factor is easily calculated and reported by a company in terms of measuring its initial stock or inventory, its production, and its waste.

Water Consumption

Freshwater is a precious commodity for business, and it is essential for all life, as we know it. Industry consumes more freshwater for business purposes than for residential or household uses. Some industrial users are more intense than others, and agribusiness leads the list of intense freshwater users. Sustainable consumption and production of commercial items will require not only production that does not pollute existing sources of precious freshwater, but also measures that reduce the intensity of its use and conserves it.

References

Conca, Ken. 2005. *Governing Water: Contentious Transnational Politics and Global Institutions Building.* Cambridge, MA: MIT Press.

Griffin, Ronald C. 2006. *Water Resources Economics: The Analysis of Scarcity, Policies and Projects.* Cambridge, MA: MIT Press.

Hannesson, Rognvaldur. 2006. *The Privatization of the Oceans.* Cambridge, MA: MIT Press.

Employee Safety and Health

One other significant indicator of a company's achievements in the field of sustainability that is often overlooked is employee safety and health. Environmental improvements that are achieved by confining waste and pollution to places where there are employees are not sustainable. Factors like indoor air pollution are often not considered by environmentalists who are looking at external consequences, but these conditions are within the domain of state and federal agencies like the Occupational Safety and Health Administrations at both state and federal levels. There are also developing international standards.

Reference

DeSombre, Elizabeth R. 2006. *Flagging Standards: Globalization and Environmental, Safety, and Labor Regulations at Sea.* Cambridge, MA: MIT Press.

TECHNOLOGY AND INFORMATION SHARING

The basic knowledge of how to develop renewable energy resources—how to produce more with less of everything—may be protected as private property under laws protecting intellectual property. Intellectual property is certain kinds of knowledge and information developed by private persons and owned by them. This property includes patented applications and copyrighted information. Businesses have sometimes invested substantial sums of money into research and development of this knowledge or its applications in practical forms. Intellectual property laws allow them to charge money for the use of their knowledge. The idea behind treating information as property in circumstances like these is to provide an incentive to creative persons, inventors, and researchers and those companies that employ them.

The problem with the basic assumptions of the intellectual property model arises when a need for the knowledge and applications is critical, but the need is unmet because of lack of money. This problem is particularly acute internationally between developing countries that are being encouraged to develop in sustainable ways and companies in the developed countries with information and technology that could facilitate sustainable development. Developing countries lack the money to pay fees for protected applications that would allow them to develop in new ways. Without that access to new technologies, these countries follow existing, unsustainable patterns of development, and our ecological decline will increase sharply as they proceed to develop in that way. Although it is in the short-term interest of businesses with technology and protected information to protect these assets as property, it is in no one's long-term interest to allow unsustainable development, especially on a massive scale contemplated by economies the size of China and India.

Companies that own valuable technology and information, and international trading organizations, could change their policies and rules in favor of greater dissemination of protected intellectual property and technologies.

Reference

Ho, Mun S., and Chris P. Nielsen. 2007. *Clearing the Air: The Health and Economic Damages of Air Pollution in China.* Cambridge, MA: MIT Press.

THE UN-GREEN ECONOMIC INITIATIVE

The UN Environmental Programme (UNEP) has called for a redesign of the global economy to invest in clean technologies and environmental sustainability of our natural resources in its Green Economy Initiative. Some have called this initiative a global Green New Deal.

Three "pillars" shape this policy initiative. First is the establishment of national and international natural services accounts. These accounts

The Lifestraw

The Lifestraw is a small, ingenious device with the power to dramatically improve the lives of millions of people without access to clean safe drinking water. The straw is actually a tube about the size of a big cigar containing patented material called PuroTech Disinfecting Resin that filters and kills bacteria and microorganisms. Lifestraw was invented by Mikkel Vestergaard Frandsen and is owned, manufactured, and marketed by Vestergaard Frandsen, a for-profit company. It would save the lives of millions of children who die from diarrhea and other waterborne illnesses each year. The device costs only $2 (U.S.) but is still well beyond the means of many poor people who are living in squalid conditions in refugee camps, urban slums, and other areas without access to free, clean drinking water.

Vestergaard Frandsen has connected its mission statement with the achievement of the Millennium Development Goals and partners with nonprofit organizations to distribute its products in developing nations.

attempt to identify a financial value for the assets that nature provides, as urged by green economists. Second, the initiative calls for encouraging green job creation by using the power of government to establish public policies encouraging such work. Third, the initiative calls for governments to establish instruments and market signals that will accelerate economic transition. UNEP will create a comprehensive assessment and tool kit including ideas for shifting subsidies and using other market-based mechanisms for change.

Five industrial sectors are the focuses of this effort. They are clean energy and clean technologies, rural energy, sustainable agriculture, ecosystem infrastructure, forests, and cities.

UNEP has also proposed the Green Economy Initiative (GEI) aimed at seizing rapidly transforming economies from inefficient dependency on fossil fuel toward clean productions and sustainable development. The GEI includes *The Green Economy Report* devoted to in-depth reporting and analysis on public policy that can accelerate this type of economic transformation. It also includes a project called the Economics of Ecosystems and Biodiversity (TEEB) that will develop ways to value accurately the benefits of ecosystems and their services. Finally, this initiative will publish the Green Jobs report describing labor trends and employment opportunities in nonpolluting and restorative sectors.

Reference

Green Economy Initiative Home Page. www.unep.org/greeneconomy.

Appendix A: Portal Web Sites

Agenda 21, www.un.org/esa/sustdev/documents/agenda21

Association for the Advancement of Sustainability in Higher Education www.aashe.org/index.php

Buckminster Fuller: Starting with the Universe and Who is Buckminster Fuller—An Introduction to Buckminster Fuller at www.bfi.org/,7/8

Center for Bhutan Studies, Gross National Happiness, www.grossnationalhappiness.com/gnhIndex/gnhIndexVariables.aspx

EPA sectors strategy program www.epa.gov/sectors/

Global Reporting Initiative www.globalreporting.org/AboutGRI/WhoWeAre/

The Green Economy Initiative, www.unep.org/greeneconomy

Green Press Initiative. www.greenpressinitiative.org/

Green suppliers network www.greensuppliers.gov/gsn/page.gsn?id=about

Greenhouse Gas Protocol Initiative, www.ghgprotocol.org/

Growing Unequal? Income Distribution and Poverty in OECD Countries, www.oecd.org

Law Office Sustainability Tools, www.earthleaders.org/olsf/office_practices

Limitations on landfill capacity for the US by state and region see soils.usda.gov/survey/geography/hurricane/index.html

Manufacturing committed to sustainability www.interfacesustainability.com/robert.html

Meadows, Donella. Leverage Points Places To Intervene in A System, www.sustainer.org/pubs/Leverage_Points.pdf

Millenium Development Goals, http://www.un.org/millenniumgoals

Millenium Ecosystem Assessment, www.millenniumassessment.org

Millennium Seed Bank Project online at www.kew.org/msbp/visit/index.htm

Our Common Future, www.un-documents.net/wced-ocf.htm

Profiles of the 2004 Blue Planet Prize Recipients, Dr. Gro Harlem Brundtland at www.af-info.or.jp/eng/honor/hot/enr-brundtland.html

Report of the World Commission on Environment and Development: Our Common Future Transmitted to the General Assembly as an Annex to document A/42/427—Development and International Co-operation: Environment

Roy Anderson www.interfaceinc.com/goals/sustainability_overview.html

TEEB—The Economics of Ecosystems and Biodiversity, www.unep.org/greeneconomy/index2.asp?id=teeb

Towards a Sustainable America: Advancing Prosperity, Opportunity, and a Healthy Environment for the 21st Century, May 1999, clinton2.nara.gov/PCSD/Publications

The Sunday Times January 11, 2009, technology.timesonline.co.uk/tol/news/tech_and_web/article5489134.ece

US EPA Brownfields and Land Revitalization www.epa.gov/swerosps/bf/index.html

World Development Indicators 2008, The World Bank, web.worldbank.org

World Resources Institute, www.earthtrends.wri.org

Worldwatch Institute, www.worldwatch.org/node/1066/print

Appendix B: The Equator Principles

A financial industry benchmark for determining, assessing and managing social and environmental risk in project financing.

PREAMBLE

Project financing, a method of funding in which the lender looks primarily to the revenues generated by a single project both as the source of repayment and as security for the exposure, plays an important role in financing development throughout the world. Project financiers may encounter social and environmental issues that are both complex and challenging, particularly with respect to projects in the emerging markets. The Equator Principles Financial Institutions (EPFIs) have consequently adopted these Principles in order to ensure that the projects we finance are developed in a manner that is socially responsible and reflect sound environmental management practices. By doing so, negative impacts on project-affected ecosystems and communities should be avoided where possible, and if these impacts are unavoidable, they should be reduced, mitigated and/or compensated for appropriately. We believe that adoption of and adherence to these Principles offers significant benefits to ourselves, our borrowers and local stakeholders through our borrowers' engagement with locally affected communities. We therefore recognise that our role as financiers affords us opportunities to promote responsible environmental stewardship and socially responsible development. As such, EPFIs will consider reviewing these Principles from time-to-time based on implementation experience, and in order to reflect ongoing learning and emerging good practice.

These Principles are intended to serve as a common baseline and framework for the implementation by each EPFI of its own internal social and environmental policies, procedures and standards related to its project financing activities. We will not provide loans to projects where the borrower will not or is unable to comply with our respective social and environmental policies and procedures that implement the Equator Principles.

SCOPE

The Principles apply to all new project financings globally with total project capital costs of US$10 million or more, and across all industry sectors. In addition, while the Principles are not intended to be applied retroactively, we will apply them to all project financings covering expansion or upgrade of an existing facility where changes in scale or scope may create significant environmental and/or social impacts, or significantly change the nature or degree of an existing impact.

The Principles also extend to project finance advisory activities. In these cases, EPFIs commit to make the client aware of the content, application and benefits of applying the Principles to the anticipated project, and request that the client communicate to the EPFI its intention to adhere to the requirements of the Principles when subsequently seeking financing.

STATEMENT OF PRINCIPLES

EPFIs will only provide loans to projects that conform to Principles 1–9 below

Principle 1: Review and Categorisation

When a project is proposed for financing, the EPFI will, as part of its internal social and environmental review and due diligence, categorise such project based on the magnitude of its potential impacts and risks in accordance with the environmental and social screening criteria of the International Finance Corporation (IFC).

Principle 2: Social and Environmental Assessment

For each project assessed as being either Category A or Category B, the borrower has conducted a Social and Environmental Assessment ("Assessment") process2 to address, as appropriate and to the EPFI's satisfaction, the relevant social and environmental impacts and risks of the proposed project (which may include, if relevant, the illustrative list of issues as found in Exhibit II). The Assessment should also propose mitigation and management measures relevant and appropriate to the nature and scale of the proposed project.

Principle 3: Applicable Social and Environmental Standards

For projects located in non-OECD countries, and those located in OECD countries not designated as High-Income, as defined by the World Bank Development Indicators Database, the Assessment will refer to the then applicable IFC Performance Standards The regulatory, permitting and public comment process requirements in High-Income OECD Countries, as defined by the World Bank Development Indicators Database, generally meet or exceed the requirements of the IFC Performance Standards (Exhibit III) and EHS Guidelines (Exhibit IV). Consequently, to avoid duplication and streamline EPFI's review of these projects, successful completion of an Assessment (or its equivalent) process under and in compliance with local or national law in High-Income OECD Countries is considered to be an acceptable substitute for the IFC Performance Standards, EHS Guidelines and further requirements as detailed in Principles 4, 5 and 6 below. For these projects, however, the EPFI still categorises and reviews the project in accordance with Principles 1 and 2 above.

The Assessment process in both cases should address compliance with relevant host country laws, regulations and permits that pertain to social and environmental matters.

Principle 4: Action Plan and Management System

For all Category A and Category B projects located in non-OECD countries, and those located in OECD countries not designated as High-Income, as defined by the World Bank Development Indicators Database, the borrower has prepared an Action Plan (AP)3 which addresses the relevant findings, and draws on the conclusions of the Assessment. The AP will describe and prioritise the actions needed to implement mitigation measures, corrective actions and monitoring measures necessary to manage the impacts and risks identified in the Assessment. Borrowers will build on, maintain or establish a Social and Environmental Management System that addresses the management of these impacts, risks, and corrective actions required to comply with applicable host country social and environmental laws and regulations, and requirements of the applicable Performance Standards and EHS Guidelines, as defined in the AP.

For projects located in High-Income OECD countries, EPFIs may require development of an Action Plan based on relevant permitting and regulatory requirements, and as defined by host-country law.

Principle 5: Consultation and Disclosure

For all Category A and, as appropriate, Category B projects located in non-OECD countries, and those located in OECD countries not designated as High-Income, as defined by the World Bank Development Indicators Database, the government, borrower or third party expert has consulted with project affected communities in a structured and culturally appropriate manner. For projects with significant adverse impacts on affected communities, the process will ensure their free, prior and informed consultation and facilitate their informed participation as a means to establish, to the satisfaction of the EPFI, whether a project has adequately incorporated affected communities' concerns.

In order to accomplish this, the Assessment documentation and AP, or non-technical summaries thereof, will be made available to the public by the borrower for a reasonable minimum period in the relevant local language and in a culturally appropriate manner. The borrower will take account of and document the process and results of the consultation, including any actions agreed resulting from the consultation. For projects with adverse social or environmental impacts, disclosure should occur early in the Assessment process and in any event before the project construction commences, and on an ongoing basis.

Principle 6: Grievance Mechanism

For all Category A and, as appropriate, Category B projects located in non-OECD countries, and those located in OECD countries not designated as High-Income, as defined by the World Bank Development Indicators Database, to ensure that consultation, disclosure and community engagement continues throughout construction and operation of the project, the borrower will, scaled to the risks and adverse impacts of the project, establish a grievance mechanism as part of the management system. This will allow the borrower to receive and facilitate resolution of concerns and grievances about the project's social and environmental performance raised by individuals or groups from among project-affected communities. The borrower will inform the affected communities about the mechanism in the course of its community engagement process and ensure that the mechanism addresses concerns promptly and transparently, in a culturally appropriate manner, and is readily accessible to all segments of the affected communities.

Principle 7: Independent Review

For all Category A projects and, as appropriate, for Category B projects, an independent social or environmental expert not directly associated with the borrower will review the Assessment, AP and consultation process documentation in order to assist EPFI's due diligence, and assess Equator Principles compliance.

Principle 8: Covenants

An important strength of the Principles is the incorporation of covenants linked to compliance. For Category A and B projects, the borrower will covenant in financing documentation:

a) to comply with all relevant host country social and environmental laws, regulations and permits in all material respects;

b) to comply with the AP (where applicable) during the construction and operation of the project in all material respects;

c) to provide periodic reports in a format agreed with EPFIs (with the frequency of these reports proportionate to the severity of impacts, or as required by law, but not less than annually), prepared by in-house staff or third party experts, that i) document compliance with the AP (where applicable), and ii) provide representation of compliance with relevant local, state and host country social and environmental laws, regulations and permits; and

d) to decommission the facilities, where applicable and appropriate, in accordance with an agreed decommissioning plan.

Where a borrower is not in compliance with its social and environmental covenants, EPFIs will work with the borrower to bring it back into compliance to the extent feasible, and if the borrower fails to re-establish compliance within an agreed grace period, EPFIs reserve the right to exercise remedies, as they consider appropriate.

Principle 9: Independent Monitoring and Reporting

To ensure ongoing monitoring and reporting over the life of the loan, EPFIs will, for all Category A projects, and as appropriate, for Category B projects, require appointment of an independent environmental and/or social expert, or require that the borrower retain qualified and experienced external experts to verify its monitoring information which would be shared with EPFIs.

Principle 10: EPFI Reporting

Each EPFI adopting the Equator Principles commits to report publicly at least annually about its Equator Principles implementation processes and experience, taking into account appropriate confidentiality considerations.

Appendix C: The Ceres Principles

The Ceres Principles are:

Protection of the Biosphere

We will reduce and make continual progress toward eliminating the release of any substance that may cause environmental damage to the air, water, or the earth or its inhabitants. We will safeguard all habitats affected by our operations and will protect open spaces and wilderness, while preserving biodiversity.

Sustainable Use of Natural Resources

We will make sustainable use of renewable natural resources, such as water, soils and forests. We will conserve non-renewable natural resources through efficient use and careful planning.

Reduction and Disposal of Wastes

We will reduce and where possible eliminate waste through source reduction and recycling. All waste will be handled and disposed of through safe and responsible methods.

Energy Conservation

We will conserve energy and improve the energy efficiency of our internal operations and of the goods and services we sell. We will make every effort to use environmentally safe and sustainable energy sources.

Risk Reduction

We will strive to minimize the environmental, health and safety risks to our employees and the communities in which we operate through safe technologies, facilities and operating procedures, and by being prepared for emergencies.

Safe Products and Services

We will reduce and where possible eliminate the use, manufacture or sale of products and services that cause environmental damage or health or safety hazards. We will inform our customers of the environmental impacts of our products or services and try to correct unsafe use.

Environmental Restoration

We will promptly and responsibly correct conditions we have caused that endanger health, safety or the environment. To the extent feasible, we will redress injuries we have caused to persons or damage we have caused to the environment and will restore the environment.

Informing the Public

We will inform in a timely manner everyone who may be affected by conditions caused by our company that might endanger health, safety or the environment. We will regularly seek advice and counsel through dialogue with persons in communities near our facilities. We will not take any action against employees for reporting dangerous incidents or conditions to management or to appropriate authorities.

Management Commitment

We will implement these Principles and sustain a process that ensures that the Board of Directors and Chief Executive Officer are fully informed about pertinent environmental issues and are fully responsible for environmental policy. In selecting our Board of Directors, we will consider demonstrated environmental commitment as a factor.

Audits and Reports

We will conduct an annual self-evaluation of our progress in implementing these Principles. We will support the timely creation of generally accepted environmental audit procedures. We will annually complete the Ceres Report, which will be made available to the public.

See www.ceres.org.

Bibliography

Ackerman, Frank and Lisa Heinzerling. 2004. *Priceless: On Knowing the Price of Everything and the Value of Nothing.* New York: New Press.

Adger, W. Neil et al., eds. 2006. *Fairness in Adaptation to Climate Change.* Cambridge, MA: MIT Press.

Aksoy, Ataman M., and John C. Beghin. 2005. *Global Agricultural Trade and Developing Countries.* Washington, DC: The World Bank.

Ali, Saleem H. 2007. *Peace Parks: Conservation and Conflict Resolution.* Cambridge, MA: MIT Press.

Allen, Jenny. 2006. *Smart Permaculture Design.* London, UK: New Holland Publishers.

Allen, Timothy F. H., Joseph A. Tainter, and Thomas W. Hoekstra. 2003. *Supply-Side Sustainability.* New York: Columbia University Press.

Allen-Gil, Susan et al. 2008. *Addressing Global Environmental Security through Innovative Educational Curricula.* New York: Springer.

Amacher, Gregory et al. 2009. *The Economics of Forest Resources.* Cambridge, MA: MIT Press.

Anderson, William. 2001. *Economics, Equity, Environment.* Washington, DC: Environmental Law Institute.

Angotti, Tom. 2008. *New York for Sale: Community Planning Confronts Global Real Estate.* Cambridge, MA: MIT Press.

Ansell, Christopher, and David Vogel. 2006. *What's the Beef? The Contested Governance of European Food Safety.* Cambridge, MA: MIT Press.

Baber, Walter F., and Robert V. Bartlett. 2005. *Deliberative Environmental Politics: Democracy and Ecological Rationality.* Cambridge, MA: MIT Press.

Bachmann, Peter, Michael Kohl, and Risto Paivinen. 1998. *Assessment of Biodiversity for Improved Forest Planning.* New York: Springer.

Bailey, Gilbert Ellis. 2008. *Vertical Farming.* Glacier National Park, MT: Kessinger Publishing.

Bailey, Robert G., and L. Ropes. 2002. *Ecoregion-Based Design for Sustainability.* New York: Springer.

Bakan, Joel. 2004. *The Corporation: The Pathological Pursuit of Profit and Power.* New York: Free Press.

Barlett, Peggy F., ed. 2005. *Urban Place: Reconnecting with the Natural World.* Cambridge, MA: MIT Press.

Barlett, Peggy F., and Geoffrey W. Chase. 2004. *Sustainability on Campus: Stories and Strategies for Change.* Cambridge, MA: MIT Press.

Barry, John, and Robyn Eckersley, eds. 2005. *The State and the Global Ecological Crisis.* Cambridge, MA: MIT Press.

Bass, Ronald E. et al. 2001. *The NEPA Book: A Step by Step Guide on How to Comply with the National Environmental Policy Act.* Point Arena, CA: Solano Press.

Beer, Tom, and Alik Ismail-Zadeh. 2002. *Risk Science and Sustainability: Science for Reduction of Risk and Sustainable Development of Society.* New York: Springer.

Bell, Simon, and Stephen Morse. 2008. *Sustainability Indicators: Measuring the Immeasurable?* London, UK: Earthscan.

Benfield, F. Kaid et al. 2001. *Solving Sprawl: Models of Smart Growth in Communities across America.* Washington, DC: Island Press.

Benyus, Janine M. 2002. *Biomimicry: Innovation Inspired by Nature.* New York: Perennial.

Bergh, J. C. van den, and M. W. Hofkes. 1998. *Theory and Implementation of Economic Models for Sustainable Development.* New York: Springer.

Binley, Dan, and Oleg Menyailo. 2004. *Tree Species Effects on Soils: Implications for Global Change.* New York: Springer.

Blewitt, John. 2008. *Community Development, Empowerment and Sustainable Development.* Devon, UK: Green Books.

Bowler, I. et al. 2002. *The Sustainability of Rural Systems: Geographical Interpretations.* New York: Springer.

Bradford, Travis. 2008. *Solar Revolution: The Economic Transformation of the Global Energy Industry.* Cambridge, MA: MIT Press.

Bramwell, Bill. 2004. *Coastal Mass Tourism: Diversification and Sustainable Development in Southern Europe.* Someset, UK: Channel View Publications.

Breitmeier, Helmut, Oran R. Young, and Michael Zurn. 2006. *Analyzing International Environmental Regimes: From Case Study to Database.* Cambridge, MA: MIT Press.

Brundtland, Gro Harlem. 1987. *Our Common Future: Report of the World Commission on Environment and Development.* New York: United Nations.

Bullard, Robert D. and Glenn S. Johnson, eds. 1997. *Just Transportation: Dismantling Race and Class Barriers in Mobility.* Gabriola Island, BC: New Society Publishers.

Capra, Fritjof. 1996. *The Web of Life: A New Scientific Understanding of Living Systems.* New York: Anchor Books.

Chichilnishy, Graciela, and Geoffrey Heal. 2000. *Environmental Markets. Equity and Efficiency.* New York: Columbia University Press.

Choi, Euiso. 2007. *Piggery Waste Management: Towards a Sustainable Future.* London, UK: IWA Publishing.

Chomsky, Noam. 2006. *Language and Mind.* Cambridge, England: Cambridge University Press.

Christensen, Julia. 2008. *Big Box Reuse.* Cambridge, MA: MIT Press.

Clapp, Jennifer, and Doris Fuchs, eds. 2009. *Corporate Power in Global Agrifood Governance.* New York: Columbia University Press.

Clapp, Jennifer, and Peter Dauvergne. 2005. *Paths to a Green World: The Political Economy of the Global Environment.* Cambridge, MA: MIT Press.

Cole, Raymond, and Richard Lorch. 2003. *Buildings, Culture and the Environment: Informing Local and Global Practices.* Hoboken, NJ: Wiley-Blackwell.

Collin, Robert W. 2006. *The Environmental Protection Agency: Cleaning up America's Act.* Westport, CT: Greenwood Press.

Collin, Robert. W. 2008. *Battleground: Environment.* Westport, CT: Greenwood Press.

Conca, Ken. 2005. *Governing Water: Contentious Transnational Politics and Global Institution Building.* Cambridge, MA: MIT Press.

Corcoran, Peter, Wlas Blaze, and E. J. Arfen. 2007. *Higher Education and the Challenge of Sustainability: Problematics, Promise, and Practice.* New York: Springer.

Costanza, Robert and Lisa Wainger. 1991. *Ecological Economics: The Science and Management of Sustainability.* New York: Columbia University Press.

Costanza, Robert, Lisa J. Graumlich, and Will Steffen, eds. 2007. *Sustainability or Collapse? An Integrated History and Future of People on Earth.* Cambridge, MA: MIT Press.

Cutter, Susan L. 1993. *Living with Risk: The Geography of Technological Hazards.* London; New York: E. Arnold.

Dauvergne, Peter. 2008. *The Shadows of Consumption: Consequences for the Global Environment.* Cambridge, MA: MIT Press.

Davenport, John, and Julia L. Davenport. 2006. *The Ecology of Transportation: Managing Mobility for the Environment.* Boca Raton, FL: Springer.

Dernbach, John, ed. 1992. *Stumbling toward Sustainability.* Washington DC: Environmental Law Institute, 45–63.

De-Shalit, Avener. 1995. *Why Posterity Matters: Environmental Policies and Future Generation.* London: Routledge.

DeSombre, Elizabeth R. 2006. *Flagging Standards: Globalization and Environmental, Safety, and Labor Regulations at Sea.* Cambridge, MA: MIT Press.

De Sousa, Christopher. 2008. *Brownfields Redevelopment and Quest for Sustainability.* Oxford, UK: Elsevier Science.

Dietz, Thomas and Paul C. Stern. 2008. *Public Participation in Environmental Assessment and Decision Making.* Washington, DC: National Academies Press.

Dillard, Jesse, Veronica Dujon, and Mary C. King. 2008. *Understanding the Social Dimension of Sustainability.* Hampshire, UK: Routledge.

Dimento, Joseph F. C. and Pamela Doughman. 2007. *Climate Change: What It Means for Us, Our Children, and Our Grandchildren.* Cambridge, MA: MIT Press.

Dixon, Tom et al., eds. 2007. *Sustainable Brownfields Regeneration: Livable Places from Problem Spaces.* Hoboken, NJ: Wiley-Blackwell.

Dobson, Andrew, and Derek Bell, eds. 2005. *Environmental Citizenship.* Cambridge, MA: MIT Press.

Durant, Robert F. 2007. *The Greening of the US Military.* Washington, DC: Georgetown University Press.

Eccleson, Charles H. 2008. *NEPA and Environmental Planning: Tools and Techniques and Approaches for Practitioners.* Boca Raton, FL: CRC Press.

Eckersley, Robyn. 2004. *The Green State: Rethinking Democracy and Sovereignty.* Cambridge, MA: MIT Press.

Egan, Michael. 2009. *Barry Commoner and the Science of Survival: The Remaking of American Environmentalism.* Cambridge, MA: MIT Press.

Ehrlich, Anne H., and John W. Birks, eds. 1990. *Hidden Dangers: Environmental Costs of Preparing for War.* San Francisco: Sierra Books.

Elkins, Paul. 2005. *Economic Growth and Environmental Sustainability.* London, UK: Routledge.

Emanual Kerry. 2007. *What We Know about Climate Change.* Cambridge, MA: MIT Press.

Emerton, Lucy et al. 2006. *Sustainable Financing of Protected Areas: A Global Review of Challenges and Options.* Washington, DC: International Union for Conservation of Nature.

Esty, Daniel C., and Andrew S. Winston. 2006. *Green to Gold: How Smart Companies Use Environmental Strategies to Innovate, Create Value, and Build Competitive Advantage.* New Haven, CT: Yale University Press.

Fairfax, Sally K. et al. 2005. *Buying Nature: The Limits of Land Acquisition as a Conservation Strategy, 1780–2004.* Cambridge, MA: MIT Press.

Felleman, John. 1997. *Deep Information: The Role of Information Policy in Environmental Sustainability.* Westport, CT: Greenwood Press.

Fiorino, Daniel J. 2006. *The New Environmental Regulation.* New York: Columbia University Press.

Flynn, James, Paul C. Slovic, and K. Gender Mertz, "Race, and Perception of Environmental Health Risks." *Journal of Risk Analysis* 14 (1994):1101–8.

Freidman, Thomas L. 2000. *The Lexus and the Olive Tree: Understanding Globalization.* Harpswell, ME: Anchor.

Freyfogle, Eric T. 2003. *The Land We Share: Private Property and the Common Good.* Washington, DC: Island Press/Shearwater Books.

Fuchs, D. A. 2003. *An Institutional Basis for Environmental Stewardship: The Structure and Quality of Property Rights.* New York: Springer.

Geisler, Charles, and Gail Daneker, eds. 1997. *Property and Values: Alternatives to Public and Private Ownership.* Washington, DC: Island Press.

Gibson, Robert B. et al. 2005. *Sustainability Assessment: Criteria and Processes.* London, UK: Earthscan.

Giudice, Fabio et al. 2006. *Product Design for the Environment: A Life Cycle Approach.* Boca Raton, FL: CRC Press.

Gore, Al. 1992. *Earth in the Balance: Ecology and the Human Spirit.* Boston, MA: Houghton Mifflin.

Gottlieb, Robert. 2007. *Reinventing Los Angeles: Nature and Community in the Global City.* Cambridge, MA: MIT Press.

Graedel, Thomas E. et al. 2005. *Greening the Industrial Facility: Perspectives, Approaches, and Tools.* New York: Springer Press.

Gray, John. 1999. *False Dawn: The Delusions of Global Capitalism.* New York: New Press.

Gray, P. M. et al. "The Music of Nature and the Nature of Music." *Science* 291 (2001):52–54.

Griffin, Ronald C. 2006. *Water Resource Economics: The Analysis of Scarcity, Policies, and Projects.* Cambridge, MA: MIT Press.

Gross, Julian. 2005. *Community Benefits Agreements: Making Development Projects Accountable.* San Francisco: Good Jobs First.

Gunn, Angus M. 2003. *Unnatural Disasters: Case Studies of Human Induced Environmental Catastrophes.* Westport, CT: Greenwood Press.

Guthman, Julie. 2004. *Agrarian Dreams: The Paradox of Organic Farming in California.* San Francisco: University of California Press.

Hannesson, Rognavaldur. 2001. *Investing for Sustainability: The Management of Mineral Wealth.* Norwell, MA: Kluwer Academic Publishers.

Hannesson, Rognvaldur. 2006. *The Privatization of the Oceans.* Cambridge, MA: MIT Press.

Harris, Jonathan Mark. 2003. *Rethinking Sustainability: Power, Knowledge, and Institutions.* Ann Arbor: University of Michigan Press.

Hart, Stuart L. 2005. *Capitalism at the Crossroads: The Unlimited Business Opportunities in Solving the World's Most Difficult Problems.* Philadelphia: Wharton School Publishing.

Hawken, Paul. 1993. *The Ecology of Commerce: A Declaration of Sustainability.* New York: HarperBusiness.

Hawken, Paul. "Natural Capitalism." *Mother Jones Magazine* March/April (1997):42.

Hawken, Paul, Amory Lovins, and L. Hunter Lovins. 1999. *Natural Capitalism: Creating the Next Industrial Revolution.* Boston: Little, Brown.

Heal, Geoffrey. 2000. *Valuing the Future: Economic Theory and Sustainability.* New York: Columbia University Press.

Heal, Geoffrey. 2008. *When Principles Pay: Corporate Social Responsibility and the Bottom Line.* New York: Columbia University Business School Publishing.

Helming, Katharina et al. 2008. *Sustainability Impact Assessment of Land Use Changes.* New York: Springer.

Hemmati, Minu et al. 2002. *Multi-Stakeholder Processes for Governance and Sustainability.* London, UK: Earthscan.

Hepworth, Adrian. 2008. *Wild Costa Rica: The Wildlife and Landscapes of Costa Rica.* Cambridge, MA: MIT Press.

Hess, David J. 2007. *Alternative Pathways in Science and Industry: Activism, Innovation, and the Environment in an Era of Globalization.* Cambridge, MA: MIT Press.

Hess, David J. 2009. *Localist Movements in a Global Economy: Sustainability, Justice, and Urban Development in the US.* Cambridge, MA: MIT Press.

Hitchcock, Darcy, and Marsha Willard. 2006. *The Business Guide to Sustainability: Practical Strategies and Tools for Organizations.* London, UK: Earthscan.

Hlavinek, Petr et al. 2009. *Risk Management of Water Supply and Sanitation Systems.* New York: Springer.

Ho, Mun S., and Chris P. Nielsen. 2007. *Clearing the Air: The Health and Economic Damages of Air Pollution in China.* Cambridge, MA: MIT Press.

Hof, John and Michael Bevers. 1998. *Spatial Optimization for Managed Ecosystems.* New York: Columbia University Press.

Hofrichter, Richard, ed. 2000. *Reclaiming the Environmental Debate: The Politics of Health in a Toxic Culture.* Cambridge, MA: MIT Press.

Hunkeler, David et al. 2008. *Environmental Life Cycle Costing.* Boca Raton, FL: CRC Press.

Jacobs, Jane. 1992. *Systems of Survival: A Dialogue on the Moral Foundations of Commerce and Politics.* New York: Vintage Books.

Jacobs, Jane. 2001. *The Nature of Economies.* New York: Random House, Vintage Books.

Janic, Milan. 2007. *The Sustainability of Air Transport: A Quantitative Analysis and Assessment.* Surrey, UK: Ashgate.

Johnson, Steven M. 2004. *Economics, Equity, and the Environment.* Washington, DC: Environmental Law Institute.

Josephson, Paul R. 2005. *Resources Under Regimes: Technology, Environment, and the State.* Cambridge, MA: Harvard University Press.

Just, Richard E., and Sinaia Netanyahu. 1998. *Conflict and Cooperation on Trans-Boundary Water Resources.* New York: Springer.

Kang, Manijit S., ed. 2007. *Agricultural and Environmental Sustainability: Considerations for the Future.* Boca Raton, FL: CRC Press.

Kassim, Tarek A., and Kenneth J. Williamson. 2005. *Environmental Impact Assessment of Recycled Wastes on Surface and Ground Waters.* New York: Springer.

Kibbel, Paul Stanton, ed. 2007. *Rivertown: Rethinking Urban Rivers.* Cambridge, MA: MIT Press.

Klijn, Frans, ed. 2007. *Ecosystem Classification for Environmental Management.* New York: Springer.

Knechtel, John. 2008. *Fuel.* Cambridge: MA: MIT Press.

Kraft, Michael E. and Sheldon Kamieniecki. 2007. *Business and Environmental Policy: Corporate Interests in the American Political System.* Cambridge, MA: MIT Press.

Leal Filho, Walter, ed. 2005. *Handbook of Sustainability Research: Environmental Education, Communication and Sustainability.* New York: Peter Land.

Lewis, Sanford J. 1993. *The Good Neighbor Handbook: A Community-Based Strategy for Sustainable Industry.* Waverly, MA: Good Neighbor Project.

Lyson, Thomas A., G. W. Stevenson, and Rick Welsh, eds. 2008. *Food and the Mid-Level Farm: Renewing an Agriculture in the Middle.* Cambridge, MA: MIT Press.

Macfarlane, Allison M. and Rodney C. Ewing. 2006. *Uncertainty Underground: Yucca Mountain and the Nation's High-Level Nuclear Waste.* Cambridge, MA: MIT Press.

Manning, Robert E. 2007. *Parks and Carrying Capacity: Commons without a Tragedy.* Washington, DC: Island Press.

McIntosh, Roderick J., Joseph A. Tainter, and Susan Keech McIntosh. 2000. *The Way the Wind Blows: Climate Change, History, and Human Action.* New York: Columbia University Press.

McNeill, J. R. 2000. *Something New under the Sun: An Environmental History of the Twentieth Century World.* New York: W. W. Norton.

Medard, Gabel, and Henry Bruner. 2003. *Global Inc.: An Atlas of the Multinational Corporation.* New York: New Press, 2–3.

Meyer, Stephen M. 2006. *The End of the Wild.* Cambridge, MA: MIT Press.

Meyers, Norman and Jennifer Kent. 2001. *Perverse Subsidies: How Tax Dollars Harm the Environment and Economy.* Washington, DC: Island Press.

Mez, Lutz. 2007. *Green Power Markets: Support Schemes, Case Studies and Perspectives.* Essex, UK: Multi-Science.

Mitchell, Ronald B., William C. Clark, David W. Cash, and Nancy M. Dickson. 2006. *Global Environmental Assessments: Information and Influence.* Cambridge, MA: MIT Press.

Montgomery, David R. 2007. *Dirt: The Erosion of Civilizations.* San Francisco: University of California Press.

Moore, Curtis. "The Impracticality and Immorality of Cost Benefit Analysis in Setting Health Related Standards." *Tulane Environmental Law Journal* 11 (1988):187, 195–98.

Moran, Emilio F., and Elinor Ostrom, eds. 2005. *Seeing the Forest and the Trees: Human-Environment Interactions in Forest Ecosystems.* Cambridge, MA: MIT Press.

Morello-Frosch, Rachel, Manuel Pastor, and James Saad. "EJ and Southern California's Riskscape: The Distribution of Air Toxics Exposures and Health Risks among Diverse Communities." *Urban Affairs Review* 36 (2001):551.

Mowforth, Martin, and Ian Munt. 1998. *Tourism and Sustainability: New Tourism in the Third World.* London: Routledge.

Myers, Nancy J. and Carolyn Raffensperger, eds. 2005. *Precautionary Tools for Reshaping Environmental Policy.* Cambridge, MA: MIT Press.

Organization for Economic Cooperation and Development. 2003. *Organic Agriculture: Sustainability, Markets, and Policy.* Wallingford, Oxfordshire, UK: CABI Publishing.

O'Riordan, Timothy, and Susanne Stoll-Kleemann. 2002. *Biodiversity, Sustainability and Human Communities: Protecting Beyond the Protected.* Cambridge, UK: Cambridge University Press.

Parr, Adrian. 2009. *Hijacking Sustainability.* Cambridge, MA. MIT Press.

Paarlberg, Robert. 2009. *Starved for Science: How Biotechnology Is Being Kept out of Africa.* Cambridge, MA: Harvard University Press.

Parry, M. L., O. F. Canziani, et al., eds. 2007. *Climate Change 2007: Impacts, Adaptation and Vulnerability. Contribution of Working Group II to the Fourth Assessment Report of the Intergovernmental Panel on Climate Change.* Cambridge, UK: Cambridge University Press.

Pirages, Dennis, and Ken Cousins. 2005. *From Resource Scarcity to Ecological Security: Exploring New Limits to Growth.* Cambridge, MA: MIT Press.

Pope, Carl, and Robert Wages. 2000. "Green Growth: Agenda for a Just Transition to a Sustainable Economy." In *The Next Agenda: Blueprint For A New Progressive Movement,* eds. Robert L. Borosage, and Roger Hickey, 249–75. Boulder, CO: Westview Press.

Princen, Thomas, Michael Maniates, and Ken Conca, eds. 2002. *Confronting Consumption.* Cambridge, MA: MIT Press.

Quinn, Daniel. 1995. *Ishmael: A Novel.* New York: Bantam/Turner Book.

Rappaport, Ann, and Sarah Hammond Creighton. 2007. *Degrees That Matter: Climate Change and the University.* Cambridge, MA: MIT Press.

Repetto, Robert C. 1986. *World Enough and Time: Successful Strategies for Resource Management.* New Haven, CT: Yale University Press.

Riddell, Robert. 2004. *Sustainable Urban Planning: Tipping the Balance.* Oxford, UK: Blackwell.

Robert, Karl-Henrik. 1995. The Ecology of Business, Bristol, UK, Schumacher Lectures, October 1995, on the Links between Social Systems, the Animal Kingdom and Our Thought Processes, 7. Online at http://www.schumacher.org.uk/transcrips/schumlec95_Bri_Theecologyofbusiness_KHRobert.pdg.

Robinson, Guy M. 2008. *Sustainable Rural Systems: Sustainable Agriculture and Rural Communities.* Aldershot, UK: Ashgate Publishing.

Robson, Mark G., and William E. Toscano. 2007. *Risk Assessment for Environmental Health.* Hoboken, NJ: Jossey-Bass.

Rogers, Heather. 2006. *Gone Tomorrow: The Hidden Life of Garbage.* New York: New Press.

Roser, Dominik et al. 2008. *Sustainable Use of Forest Biomass for Energy.* New York: Springer.

Sabatier, Paul et al., eds. 2005. *Swimming Upstream: Collaborative Approaches to Watershed Management.* Cambridge, MA: MIT Press.

Sachs, Jeffery. 2006. *The End of Poverty: Economic Possibilities for Our Time.* New York: Penguin Press.

Savitz, Andrew W, and Karl Webber. 2006. *The Triple Bottom Line: How Today's Best Run Companies Are Achieving Economic, Social, and Environmental Success.* Hoboken, NJ: Jossey-Bass.

Schafer, Andreas et al. 2009. *Transportation in a Climate Constrained World.* Cambridge, MA: MIT Press.

Schaltegger, Stefan et al. 2006. *Sustainability Accounting and Reporting.* New York: Springer.

Seliger, Gunther. 2007. *Sustainability in Manufacturing: Recovery of Resources in Product and Material Cycles.* New York: Springer.

Slobodchikoff, C. N. 2002. "Cognition and Communication in Prairie Dogs." In *The Cognitive Animal: Empirical and Theoretical Perspectives on Animal Cognition,* eds. Marc Bekoff, Colin Allen, and Gordon M. Burghard, 257–64. Cambridge, MA: MIT Press.

Smil, Vaclav. 2008. *Global Catastrophes and Trends: The Last 50 Years.* Cambridge, MA: MIT Press.

Smith, Keith. 2004. *Environmental Hazards: Assessing Risks and Reducing Disasters.* Andover, UK: Routledge.

Spellman, Frank R. 2007. *Environmental Management of Concentrated Animal Feeding Operations.* Boca Raton, FL: CRC Press.

Stein, Bruce A. et al., eds., 2000. *Our Precious Heritage: The Status of Biodiversity in the United States.* New York: Oxford University Press.

Stern, Alissa J. 2000. *The Process of Business/Environmental Collaborations: Partnering for Sustainability.* Westport, CT: Greenwood Publishing.

Svedin, Uno Britt, and Hägerhäll Aniansson. 2002. *Sustainability, Local Democracy, and the Future: The Swedish Model.* Emeryville, CA: Kluwer Academic Publishers.

Tolley, Rodney. 2003. *Sustainable Transport: Planning for Walking and Cycling in Urban Environments.* Boca Raton, FL: CRC Publishing.

Townsend, Amy K. 2006. *Green Business: A Five Part Model for Creating an Environmentally Responsible Company.* Atglen, PA: Schiffer.

Troesken, Werner. 2008. *The Great Lead Water Pipe Disaster.* Cambridge, MA: MIT Press.

United Nations. 2008. *Delivering on the Global Partnership for Achieving the Millennium Development Goals MDG Gap Task Force Report.* New York: United Nations. Online at www.un.org/esa/policy/mdggap.

United Nations Environmental Program. 2007. *Life Cycle Management: A Business Guide to Sustainability.* New York: UN Publications.

United Nations Environmental Program. 2008. *Climate Change and Tourism: Responding to Global Challenges.* New York: United nations Publications.

Volk, Tyler. 2008. *CO2 Rising: The World's Greatest Environmental Challenge.* Cambridge, MA: MIT Press.

Waltner-Toews, David, James J. Kay, and Nina-Marie E. Lister. 2008. *The Ecosystem Approach: Complexity, Uncertainty, and Managing for Sustainability.* New York: Columbia University Press.

Warner, Douglass Keith. 2007. *Agroecology in Action: Extending Alternative Agriculture through Social Networks.* Cambridge, MA: MIT Press.

Webster, D. G. 2008. *Adaptive Governance: The Dynamics of Atlantic Fisheries Management.* Cambridge, MA: MIT Press.

Whiteside, Kerry H. 2006. *Precautionary Politics: Principle and Practice in Confronting Environmental Risk.* Cambridge, MA: MIT Press.

Widyawati, Diah. 2008. *Environmental Audit and Compliance: The Role of Audit Policies in Inducing Audit Adoption and Compliance Behavior.* Saarbrücken Germany: VDM Verlag.

World Bank. 2008. *The Global Monitoring Report.* Washington, DC: World Bank.

Yanful, Earnest K. 2000. *Appropriate Technologies for Environmental Protection in the Developing World.* Boca Raton, FL: Springer.

Yang, Zin. 2008. *Strategic Bargaining and Cooperation in Greenhouse Gas Mitigations: An Integrated Modeling Approach.* Cambridge, MA: MIT Press.

Yudelson, Jerry. 2008. *Green Building through Integrated Design.* New York: McGraw Hill Professional.

Zarin, Daniel. 2004. *Working Forests in the Neotropics: Conservation through Sustainable Management?* New York: Columbia University Press.

Zovanayi, Gabor. 1998. *Growth Management for a Sustainable Future: Ecological Sustainability as the New Growth Management for the 21st Century.* Westport, CT: Greenwood Publishing.

Index

Boldface numbers refer to volume numbers. A key appears on all verso pages. Page numbers with a *f* following them indicate a figure.

1: Environment and Ecology **2:** Business and Economics **3:** Equity and Fairness

1: Environment and Ecology
2: Business and Economics
3: Equity and Fairness

1: Environment and Ecology
2: Business and Economics
3: Equity and Fairness

1: Environment and Ecology
2: Business and Economics
3: Equity and Fairness

1: Environment and Ecology
2: Business and Economics
3: Equity and Fairness

1: Environment and Ecology
2: Business and Economics
3: Equity and Fairness

1: Environment and Ecology
2: Business and Economics
3: Equity and Fairness

ABOUT THE AUTHORS

ROBIN MORRIS COLLIN, professor of law, Willamette College of Law. Professor Morris Collin has taught at McGeorge School of Law, Tulane School of Law, Pepperdine Law School, Washington and Lee School of Law, University of Oregon School of Law, and Willamette School of Law. She has numerous publications in the area of sustainability and holds the David Brower Lifetime Achievement Award. In April 2009, she was awarded the *Judith Ramaley Faculty Award for Civic Engagement in Sustainability,* Oregon Campus Compact. Professor Collin was the first law professor to teach sustainability in the United States in 1993 and has taught it ever since. She has served as an advisor to state and federal environmental agencies. She has also litigated court cases and provided legislative testimony on many important environmental issues. She is currently working with the Oregon State Bar Association to find ways to integrate sustainability into legal practice. She is also appointed to the Oregon Environmental Justice Advisory Group.

ROBERT WILLIAM COLLIN is the senior research scholar at the Center for Sustainable Communities at Willamette University. He has been a professor of law, planning, and of social work, teaching at the University of Auckland, New Zealand; the University of Virginia Department of Urban and Environmental Studies; the University of Oregon's Department of Environmental Studies; Cleveland State University Department of Social Work; Jackson State University Department of Urban and Regional Planning; Lewis and Clark College of Law; and Willamette University College of Law. He has published many articles, book chapters, and book reviews. He has served as an advisor to state and federal environmental agencies and currently serves as chair of the Oregon Environmental Justice Advisory Group. His last two books are *The US Environmental Protection Agency: Cleaning up America's Act,* and *Battleground: Environment* (2 volumes).